Charles Davies

**Key to Davies' Bourdon,**

With many Additional Examples, Illustrating the Algebraic Analysis

Charles Davies

**Key to Davies' Bourdon,**
*With many Additional Examples, Illustrating the Algebraic Analysis*

ISBN/EAN: 9783744696807

Printed in Europe, USA, Canada, Australia, Japan

Cover: Foto ©Paul-Georg Meister /pixelio.de

More available books at **www.hansebooks.com**

# KEY

TO

# DAVIES' BOURDON,

WITH

MANY ADDITIONAL EXAMPLES, ILLUSTRATING
THE ALGEBRAIC ANALYSIS.

By CHARLES DAVIES, LL.D.,
AUTHOR OF A FULL COURSE OF MATHEMATICS.

A. S. BARNES & COMPANY,
NEW YORK, CHICAGO AND NEW ORLEANS.
1876.

# DAVIES' COURSE OF MATHEMATICS.

## IN THREE PARTS.

### I. COMMON SCHOOL COURSE.

**DAVIES' PRIMARY ARITHMETIC.**—The fundamental principles displayed in Object Lessons.

**DAVIES' INTELLECTUAL ARITHMETIC.**—Referring all operations to the unit 1 as the only tangible basis for logical development.

**DAVIES' ELEMENTS OF WRITTEN ARITHMETIC.**—A practical introduction to the whole subject. Theory subordinated to Practice.

**DAVIES' PRACTICAL ARITHMETIC.**—A combination of Theory and Practice, clear, exact, brief, and comprehensive.

### II. ACADEMIC COURSE.

**DAVIES' UNIVERSITY ARITHMETIC.**—Treating the subject exhaustively as a *science*, in a logical series of connected propositions.

**DAVIES' ELEMENTARY ALGEBRA.**—A connecting link, conducting the pupil easily from arithmetical processes to abstract analysis.

**DAVIES' UNIVERSITY ALGEBRA.**—For institutions desiring a more complete but not the fullest course in pure Algebra.

**DAVIES' PRACTICAL MATHEMATICS.**—The science practically applied to the useful arts, as Drawing, Architecture, Surveying, Mechanics, etc.

**DAVIES' ELEMENTARY GEOMETRY.**—The important principles in simple form, but with all the exactness of rigorous reasoning.

**DAVIES' ELEMENTS OF SURVEYING.**—Re-written in 1870. A simple and full presentation for Instruction and Practice.

### III. COLLEGIATE COURSE.

**DAVIES' BOURDON'S ALGEBRA.**—Embracing Sturm's Theorem, and a most exhaustive course. Re-written, in 1873.

**DAVIES' UNIVERSITY ALGEBRA.**—A shorter course than Bourdon, for Institutions having less time to give the subject.

**DAVIES' LEGENDRE'S GEOMETRY.**—A standard work in this country and in Europe.

**DAVIES' ANALYTICAL GEOMETRY.**—A full course of Analysis, embracing the applications to surfaces of the second order.

**DAVIES' DIFFERENTIAL AND INTEGRAL CALCULUS**, on the basis of Continuous Quantity and Consecutive Differences.

**DAVIES' ANALYTICAL GEOMETRY AND CALCULUS.**—The shorter treatises, combined in one volume.

**DAVIES' DESCRIPTIVE GEOMETRY.**—With application to Spherical Trigonometry, Spherical Projections, and Warped Surfaces.

**DAVIES' SHADES, SHADOWS, AND PERSPECTIVE.**—A succinct exposition of the mathematical principles involved.

**DAVIES' NATURE AND UTILITY OF MATHEMATICS**, Logically considered.

**DAVIES AND PECK'S MATHEMATICAL DICTIONARY**, or Cyclopedia of Mathematics.

---

Entered according to Act of Congress, in the year 1873, by
CHARLES DAVIES,
In the Office of the Librarian of Congress, at Washington.

# PREFACE.

A WIDE difference of opinion is known to exist among teachers in regard to the value of a Key to any mathematical work, and it is perhaps yet undecided whether a Key is a help or a hindrance.

If a Key is designed to supersede the necessity of investigation and labor on the part of the teacher; to present to his mind every combination of thought which ought to be suggested by a problem, and to permit him to float sluggishly along the current of ideas developed by the author, it would certainly do great harm, and should be excluded from every school.

If, on the contrary, a Key is so constructed as to suggest ideas, both in regard to particular questions and general science, which the Text-book might not impart; if it develops methods of solution too particular or too elaborate to find a place in the text; if it is mainly designed to lessen the *mechanical labor of teaching*, rather than the labor of study and investigation; it may, in the hands of a good teacher, prove a valuable auxiliary.

The KEY TO BOURDON is intended to answer, precisely, this

end. The principles developed in the text are explained and illustrated by means of numerous examples, and these are all wrought in the Key by methods which accord with and make evident the principles themselves. The Key, therefore, not only explains the various questions, but is a commentary on the text itself.

Nothing is more gratifying to an ambitious teacher than to push forward the investigations of his pupils beyond the limits of the text book. To aid him in an undertaking so useful to himself and to them, an Appendix has been added, containing a copious collection of Practical Examples. Many of the solutions are quite curious and instructive; and taken in connection with those embraced in the Text, form a full and complete system of Algebraic Analysis.

The many letters which I have received from Teachers and Pupils, in regard to the best solutions of new questions, have suggested the desirableness of furnishing, in the present work, those which have been most approved. They are a collection of problems that have special values, and their solutions may be studied with great profit by every one seeking mathematical knowledge.

FISHKILL LANDING,
 *July, 1873.*

# INTRODUCTION.

## ALGEBRA.

1. On an analysis of the subject of Algebra, we think it will appear that the subject itself presents no serious difficulties, and that most of the embarrassment which is experienced by the pupil in gaining a knowledge of its principles, as well as in their applications, arises from not attending sufficiently to the *language* or *signs* of the thoughts which are combined in the reasonings. At the hazard, therefore, of being a little diffuse, I shall begin with the very elements of the algebraic language, and explain, with much minuteness, the exact signification of the characters that stand for the quantities which are the subjects of the analysis; and also of those signs which indicate the several operations to be performed on the quantities.

<small>Algebra.
Difficulties
How overcome.
Language.
Characters which represent quantity
Signs.</small>

2. The quantities which are the subjects of the algebraic analysis may be divided into two classes: those which are known or given, and those which are unknown or sought. The known are uniformly represented by the first letters of the alphabet, $a$, $b$, $c$, $d$, &c.; and the unknown by the final letters, $x$, $y$, $z$, $v$, $w$, &c.

<small>Quantities.
How divided.
How represented.</small>

# INTRODUCTION.

*May be increased or diminished.* Quantity is susceptible of being increased, diminished, and measured; and there are six operations *Five operations.* which can be performed upon a quantity that will give results differing from the quantity itself, viz.:

*First*      1st. To add it to itself or to some other quantity;
*Second*    2d. To subtract some other quantity from it;
*Third.*     3d. To multiply it by a number;
*Fourth.*    4th. To divide it;
*Fifth.*      5th. To raise it to any power;
             6th. To extract a root of it.

*Exception.* The cases in which the multiplier or divisor is 1, are of course excepted; as also the case where a root is to be extracted of 1.

*Signs.*   4. The six signs which denote these operations *Elements of the Algebraic language.* are too well known to be repeated here. These, with the signs of equality and inequality, the letters of the alphabet and the figures which are employed, make up *Its words and phrases:* the elements of the algebraic language. The words and phrases of the algebraic, like those of every *How interpreted.* other language, are to be taken in connection with each other, and are not to be interpreted as separate and isolated symbols.

*Symbols of quantity.*   5. The symbols of quantity are designed to represent quantity in general, whether abstract or concrete, *General.* whether known or unknown; and the signs which indicate the operations to be performed on the quanti- *Examples.* ties are to be interpreted in a sense equally general. When the sign plus is written, it indicates that the *Signs plus and minus.* quantity before which it is placed is to be added to some other quantity: and the sign minus implies the existence of a minuend, from which the subtrahend is to be taken. One thing should be observed in regard to the signs which indicate the operations that are to be performed on quantities, viz.: *they do not at all affect or change the nature of the quantity before or after which they are written, but merely indicate what is to be done with the quantity.* In Algebra, for example, the minus sign merely indicates that the quantity before which it is written is to be subtracted from some other quantity; and in Analytical Geometry, that the line before which it falls is estimated in a contrary direction from that in which it would have been reckoned, had it had the sign plus; but in neither case is the *nature* of the quantity itself different from what it would have been had the sign been plus. *[margin: Signs have no effect on the nature of a quantity. Examples: In Algebra. In Analytical Geometry.]*

The interpretation of the language of Algebra is the first thing to which the attention of a pupil should be directed; and he should be drilled on the meaning and import of the symbols, until their significations and uses are as familiar as the sounds and combinations of the letters of the alphabet. *[margin: Interpretation of the language: Its necessity.]*

6. Beginning with the elements of the language, let any number or quantity be designated by the letter $a$, and let it be required to add this letter to itself and find the result or sum. The addition will be expressed by *[margin: Elements explained]*

$$a + a = \text{the sum.}$$

But how is the sum to be expressed? By simply regarding $a$ as *one a*, or $1a$, and then observing that one $a$ and one $a$, make *two a's* or $2a$: hence, *[margin: Signification]*

8  INTRODUCTION.

$$a + a = 2a;$$

and thus we place a figure before a letter to indicate how many times it is taken. Such figure is called a *Co-efficient*.

**Co-efficient**

**Product:**
7. The product of several numbers is indicated by the sign of multiplication, or by simply writing the letters which represent the numbers by the side of each other. Thus,

**how indicated**
$$a \times b \times c \times d \times f, \text{ or } abcdf,$$

**Factor.**
indicates the continued product of $a$, $b$, $c$, $d$, and $f$, and each letter is called a factor of the product: hence, a *factor* of a product is one of the multipliers which produce it. Any figure, as 5, written before a product, as

$$5abcdf,$$

**Co-efficient of a product.**
is the co-efficient of the product, and shows that the product is taken 5 times.

**Equal factors: what the product becomes.**
8. If the numbers represented by $a$, $b$, $c$, $d$, and $f$, were equal to each other, they would each be represented by a single letter $a$, and the product would then become

**How expressed:**
$$a \times a \times a \times a \times a = a^5;$$

that is, we indicate the product of several equal factors by simply writing the letter once and placing a figure above and a little at the right of it, to indicate how many times it is taken as a factor. The figure so written is called an *exponent*. Hence, *an exponent is a simple form of language to point out how many equal factors are employed.*

Exponent: where written.

9. The division of one quantity by another is indicated by simply writing the divisor below the dividend, after the manner of a fraction; by placing it on the right of the dividend with a horizontal line and two dots between them; or by placing it on the right with a vertical line between them: thus either form of expression:

Division: how expressed

$$\frac{a}{b}, \quad b \div a, \quad \text{or} \quad b \,|\, a,$$

Three forms.

indicates the division of $b$ by $a$.

10. The extraction of a root is indicated by the sign $\sqrt{\ }$. This sign, when used by itself indicates the lowest root, viz., the square root. If any other root is to be extracted, as the third, fourth, fifth, &c., the figure marking the degree of the root is written above and at the left of the sign; as,

Roots: how indicated<br>Index; where written

$$\sqrt[3]{\ } \text{ cube root}, \quad \sqrt[4]{\ } \text{ fourth root}, \&c.$$

The figure so written, is called the *Index* of the root.

We have thus given the very simple and general language by which we indicate each of the five operations that may be performed on an algebraic quantity, and *every process in Algebra* involves one or other of these operations.

Language for the five operations

## INTRODUCTION.

### MINUS SIGN.

*Algebraic language:*

*how divided.*

11. The algebraic symbols are divided into two classes entirely distinct from each other—viz., the letters that are used to designate the quantities which are the subjects of the science, and the signs which are employed to indicate certain operations to be performed on those quantities. We have seen that all the algebraic processes are comprised under addition, subtraction, multiplication, division, and the extraction of roots; and it is plain, that the *nature* of a quantity is not at all changed by prefixing to it the sign which indicates either of these operations. The quantity denoted by the letter $a$, for example, is the same, in *every respect*, whatever sign may be prefixed to it; that is, whether it be added to another quantity, subtracted from it, whether multiplied or divided by any number, or whether we extract the square or cube or any other root of it. The algebraic signs, therefore, must be regarded merely as indicating *operations* to be performed on quantity, and not as affecting the *nature* of the quantities to which they may be prefixed. We say, indeed, that quantities are plus and minus, but this is an abbreviated language to express that they are to be added or subtracted.

*Algebraic processes:*

*their number.*

*Do not change the nature of the quantities.*

*Algebraic signs: how regarded.*

*Plus and Minus.*

*Principles of the science. From what deduced.*

*Example.*

12. In Algebra, as in Arithmetic and Geometry all the principles of the science are deduced from the definitions and axioms; and the rules for performing the operations are but directions framed in conformity to such principles. Having, for example, fixed by definition, the power of the minus sign, viz., that any

quantity before which it is written, shall be regarded as to be subtracted from another quantity, we wish to discover the process of performing that subtraction, so as to deduce therefrom a general *formula*, from which we can frame a rule applicable to all similar cases.

*What we wish to discover*

## SUBTRACTION.

13. Let it be required, for example, to subtract from $b$ the difference between $a$ and $c$. Now, having written the letters, with their proper signs, the language of Algebra expresses that it is the *difference* only between $a$ and $c$, which is to be taken from $b$; and if this difference were known, we could make the subtraction at once. But the nature and generality of the algebraic symbols, enable us to *indicate operations*, merely, and we cannot in general make reductions until we come to the final result. In what general way, therefore, can we indicate the true difference?

*Subtraction.*

$$\begin{array}{|l} b \\ a - c \end{array}$$

*Process.*

*Difference.*

*Operations indicated.*

If we indicate the subtraction of $a$ from $b$, we have $b - a$; but then we have taken away too much from $b$ by the number of units in $c$; for it was not $a$, but the *difference* between $a$ and $c$ that was to be subtracted from $b$. Having taken away *too much*, the remainder is *too small* by $c$: hence, if $c$ be added, the true remainder will be expressed by $b - a + c$.

$$\begin{array}{|l} b - a \\ b - a + c \end{array}$$

*Final formula.*

Now, by analyzing this result, we see that the sign of every term of the subtrahend has been changed; and what has been shown with respect to these quan-

*Analysis of the result.*

*Generalization.* tities is equally true of all others standing in the same relation: hence, we have the following general rule for the subtraction of algebraic quantities:

*Rule.* *Change the sign of every term of the subtrahend, or conceive it to be changed, and then unite the quantities as in addition.*

## MULTIPLICATION.

*Multiplication.* 14. Let us now consider the case of multiplication, and let it be required to multiply $a - b$ by $c$. The algebraic language expresses that the

*Signification of the language.* difference between $a$ and $b$ is to be taken as many times as there are units in $c$. If we knew this differ-

$$\begin{array}{r}a - b \\ c \\ \hline ac - bc\end{array}$$

ence, we could at once perform the multiplication.

*Process.* But by what general process is it to be performed without finding that difference? If we take $a$, $c$ times, the product will be $ac$; but as it was only the *difference* between $a$ and $b$, that was to be multiplied by $c$,

*Its nature.* this product $ac$ will be too great by $b$ taken $c$ times; that is, the true product will be expressed by $ac - bc$: hence, we see, that,

*Principle for the signs.* *If a quantity having a plus sign be multiplied by another quantity having also a plus sign, the sign of the product will be plus; and if a quantity having a minus sign be multiplied by a quantity having a plus sign, the sign of the product will be minus.*

*General case:* 15. Let us now take the most general case, viz., that in which it is required to multipy $a - b$ by $c - d$

## INTRODUCTION. 13

Let us again observe that the algebraic language denotes that $a-b$ is to be taken as many times as there are units in $c-d$; and if these two differences were known, their product would at once form the product required.

$$\begin{array}{r} a-b \\ c-d \\ \hline ac-bc \\ -ad+bd \\ \hline ac-bc-ad+bd \end{array}$$

Its form.

First: let us take $a-b$ as many times as there are units in $c$; this product, from what has already been shown, is equal to $ac-bc$. But since the multiplier is not $c$, but $c-d$, it follows that this product is too large by $a-b$ taken $d$ times; that is, by $ad-bd$: hence, the first product diminished by this last, will give the true product. But, by the rule for subtraction, this difference is found by changing the signs of the subtrahend, and then uniting all the terms as in addition: hence, the true product is expressed by $ac-bc-ad+bd$.

First step.

Second step:

How taken.

By analyzing this result, and employing an abbreviated language, we have the following general principle to which the signs conform in multiplication, viz.:

Analysis of the result.

*Plus multiplied by plus gives plus in the product; plus multiplied by minus gives minus; minus multiplied by plus gives minus; and minus multiplied by minus gives plus in the product.*

General Principle.

16. The remark is often made by pupils that the above reasoning appears very satisfactory so long as the quantities are presented under the above form; but why will $-5$ multiplied by $-7$ give plus $bd$?

Remark.

Particular case.

How can the product of two negative quantities *standing alone* be plus?

*Minus sign.* In the first place, the minus sign being prefixed to $b$ and $d$, shows that in an *algebraic sense* they do not *Its interpretation.* stand by themselves, but are connected with other quantities; and if they are not so connected, the minus sign makes no difference; for, it in no case affects the quantity, but merely points out a connection with other quantities. Besides, the product determined above, being independent of any particular value attributed *Form of the product: must be true for quantities of any value.* to the letters $a$, $b$, $c$, and $d$, must be of such a form as to be true for all values; and hence for the case in which $a$ and $c$ are each equal to zero. Making this supposition, the product reduces to the form of $+bd$. *Signs in division.* The rules for the signs in division are readily deduced from the definition of division, and the principles already laid down.

## ZERO AND INFINITY.

*Zero and Infinity.* 17. The terms zero and infinity have given rise to much discussion, and been regarded as presenting difficulties not easily removed. It may not be easy to frame a form of language that shall convey to a mind, *Ideas not abstruse.* but little versed in mathematical science, the precise ideas which these terms are designed to express; but we are unwilling to suppose that the ideas themselves are beyond the grasp of an ordinary intellect. The terms are used to designate the *two limits of Space and Number.*

18. Assuming any two points in space, and joining

them by a straight line, the distance between the points will be truly indicated by the length of this line, and this length may be expressed numerically by the number of times which the line contains a known unit. If now, the points are made to approach each other, the length of the line will diminish as the points come nearer and nearer together, until at length, when the two points become one, the length of the line will disappear, having attained its *limit*, which is called ʟero. If, on the contrary, the points recede from each other, the length of the line joining them will continually increase; but so long as the length of the line can be expressed in terms of a known unit of measure, it is not infinite. But, if we suppose the points removed, so that any known unit of measure would occupy no *appreciable portion* of the line, then the length of the line is said to be *Infinite*.

*Illustration, showing the meaning of the term Zero.*

*Illustration, showing the meaning of the term Infinity.*

19. Assuming one as the unit of number, and admitting the self-evident truth that it may be increased or diminished, we shall have no difficulty in understanding the import of the terms zero and infinity, as applied to number. For, if we suppose the unit one to be continually diminished, by division or otherwise, the fractional units thus arising will be less and less, and in proportion as we continue the divisions, they will continue to diminish. Now, the limit or boundary to which these very small fractions approach, is called Zero, or nothing. So long as the fractional number forms an appreciable part of one, it is not zero, but a finite fraction; and the term zero is only

*The terms Zero and Infinity applied to numbers.*

*Illustration.*

*Zero:*

applicable to that which forms no appreciable part of the standard.

*Illustration.* If, on the other hand, we suppose a number to be continually increased, the relation of this number to the unit will be constantly changing. So long as the number can be expressed in terms of the unit one, it is *Infinity;* finite, and not infinite; but when the unit one forms no appreciable part of the number, the term *infinite* is used to express that state of value, or rather, that limit of value.

*The terms, how employed.* 20. The terms zero and infinity are therefore employed to designate the limits to which decreasing and increasing quantities may be made to approach nearer *Are limits.* than any assignable quantity; but these limits cannot be compared, in respect to magnitude, with any known standard, so as to give a finite ratio.

*Why limits?* 21. It may, perhaps, appear somewhat paradoxical, that zero and infinity should be defined as "the limits of number and space" when they are in themselves not measurable. But a limit is that "which sets bounds *Definition of a limit.* to, or circumscribes;" and as all finite space and finite number (and such only are implied by the terms Space *Of Space and Number.* and Number), are contained between zero and infinity, we employ these terms to designate the limits of Number and Space.

## OF THE EQUATION.

*Subject of equations: how divided.* 22. The subject of equations is divided into two parts. The first, consists in finding the equation; that *First part:* is, in the process of expressing the relations existing

INTRODUCTION. 17

between the quantities considered, by means of the algebraic symbols and formula. This is called the Statement of the proposition. The second is purely deductive, and consists, in Algebra, in what is called the solution of the equation, or finding the value of the unknown quantity; and in the other branches of analysis, it consists in the discussion of the equation; that is, in the drawing out from the equation every proposition which it is capable of expressing. *Statement. Second part. Solution. Discussion of an equation*

23. Making the statement, or finding the equation, is merely analyzing the problem, and expressing its elements and their relations in the language of analysis. It is, in truth, collating the facts, noting their bearing and connection, and inferring some general law or principle which leads to the formation of an equation. *Statement: what it is.*

The condition of equality between two quantities is expressed by the sign of equality, which is placed between them. The quantity on the left of the sign of equality is called the first member, and that on the right, the second member of the equation. Hence, an equation is merely a proposition expressed algebraically, in which equality is predicated of one quantity as compared with another. It is the great formula of Algebra. *Equality of two quantities: How expressed. 1st member. 2d member. Proposition.*

24. Every quantity is either abstract or concrete: hence, an equation, which is a general formula for expressing equality, must be either abstract or concrete. *Abstract. Concrete.*

2

## INTRODUCTION.

**Abstract equation.**

An abstract equation expresses merely the relation of equality between two abstract quantities: thus,

$$a + b = x,$$

is an abstract equation, if no unit of value be assigned to either member; for, until that be done the abstract unit one is understood, and the formula merely expresses that the sum of $a$ and $b$ is equal to $x$, and is true, equally, of all quantities.

**Concrete equation.**

But if we assign a concrete unit of value, that is, say that $a$ and $b$ shall each denote so many pounds weight, or so many feet or yards of length, $x$ will be of the same denomination, and the equation will become concrete or denominate.

**Five operations may be performed.**

25. We have seen that there are five operations which may be performed on an algebraic quantity (Art. 3). We assume, as an axiom, that if the same operation, under either of these processes, be performed on both members of an equation, the equality of the members will not be changed. Hence, we have the five following

**Axioms.**

### AXIOMS.

**First.**

1. If equal quantities be added to both members of an equation, the equality of the members will not be destroyed.

**Second.**

2. If equal quantities be subtracted from both members of an equation, the equality will not be destroyed.

**Third.**

3. If both members of an equation be multiplied by the same number, the equality will not be destroyed

4. If both members of an equation be divided by the same number, the equality will not be destroyed. *Fourth.*

5. If both members of an equation be raised to the same power, the equality of the members will not be destroyed. *Fifth.*

6. If the same root of both members of an equation be extracted, the equality of the members will not be destroyed. *Sixth.*

Every operation performed on an equation will fall under one or other of these axioms, and they afford the means of solving all equations which admit of solution. *Use of axioms.*

**26.** Two quantities are said to be *equal*, when each contains the same unit an equal number of times. Hence, the term *equal* applies to *measures*, and has the same signification in Arithmetic, in Algebra, and in Geometry. If, in Geometry, two figures can be applied to each other, so as to coincide or fill the same space, they are said to be *equal in all their parts*. *Equality defined.* *Equal in all parts.*

**27.** We have thus pointed out some of the marked characteristics of Algebra. In Algebra, the quantities, about which the science is conversant, are divided, as has been already remarked, into known and unknown, and the connections between them, expressed by the equation, afford the means of tracing out further relations, and of finding the values of the unknown quantities in terms of the known. *Classes of quantities in Algebra.*

## INTRODUCTION.

### SUGGESTIONS FOR THOSE WHO TEACH ALGEBRA.

*Letters are but mere symbols.*
1. Be careful to explain that the letters employed, are the mere symbols of quantity. That of and in themselves, they have no meaning or signification whatever, but are used merely as the signs or representatives of such quantities as they may be employed to denote.

*Signs indicate operations.*
2. Be careful to explain that the signs which are used are employed merely for the purpose of indicating the five operations which may be performed on quantity; and that they indicate operations merely, without at all affecting the nature of the quantities before which they are placed.

*Letters and signs elements of language.*
3. Explain that the letters and signs are the elements of the algebraic language, and that the language itself arises from the combination of these elements.

*Algebraic formula.*
4. Explain that the finding of an algebraic formula is but the translation of certain ideas, first expressed in our common language, into the language of Algebra;

*Its interpretation.*
and that the interpretation of an algebraic formula is merely translating its various significations into common language.

*Language.*
5. Let the language of Algebra be carefully studied, so that its construction and significations may be clearly apprehended.

*Co-efficient.*
*Exponent.*
6. Let the difference between a co-efficient and an exponent be carefully noted, and the office of each often explained; and illustrate frequently the signification of the language by attributing numerical values to letters in various algebraic expressions.

7. Point out often the characteristics of similar and

# INTRODUCTION. 21

dissimilar quantities, and explain which may be incorporated and which cannot. *Similar quantities.*

8. Explain the power of the minus sign, as shown in the four ground rules, but very particularly as it is illustrated in subtraction and multiplication. *Minus sign.*

9. Point out and illustrate the correspondence between the four ground rules of Arithmetic and Algebra; and impress the fact, that their differences, wherever they appear, arise merely from differences in notation and language: the principles which govern the operations being the same in both. *Arithmetic and Algebra compared.*

10. Explain with great minuteness and particularity, all the characteristic properties of the equation; the manner of forming it; the different kinds of quantity which enter into its composition; its examination or discussion; and the different methods of elimination. *Equation. Its properties.*

11. In the equation of the second degree, be careful to dwell on the four forms which embrace all the cases, and illustrate by many examples that every equation of the second degree may be reduced to one or other of them. Explain very particularly the meaning of the term root; and then show, why every equation of the first degree has one, and every equation of the second degree two. Dwell on the properties of these roots in the equation of the second degree. Show why their sum, in all the forms, is equal to the co-efficient of the second term, taken with a contrary sign; and why their product is equal to the absolute term with a contrary sign. Explain when and why the roots are imaginary. *Equation of the second degree. Its form. Its roots. Their sum. Their product.*

*General Principles:*

*Should be explained.*

*They lead to general laws.*

12. In fine, remember that every operation and rule is based on a principle of science, and that an intelligible reason may be given for it. Find that reason, and impress it on the mind of your pupil in plain and simple language, and by familiar and appropriate illustrations. You will thus impress right habits of investigation and study, and he will grow in knowledge. The broad field of analytical investigation will be opened to his intellectual vision, and he will have made the first steps in that sublime science which discovers the laws of nature in their most secret hiding-places, and follows them, as they reach out, in omnipotent power, to control the motions of matter through the entire regions of occupied space.

(See Davies' Nature and Utility of Mathematics, **Article Algebra**).

# KEY.

## EQUATIONS OF THE FIRST DEGREE.

1. Given $\dfrac{5x}{12} - \dfrac{4x}{3} - 13 = \dfrac{7}{8} - \dfrac{13x}{6}$.

### VERIFICATION.

$$\dfrac{5 \times 11.1}{12} - \dfrac{4 \times 11.1}{3} - 13 = \dfrac{7}{8} - \dfrac{13 \times 11.1}{6}.$$

Multiply by 24, least common multiple,

$10 \times 11.1 - 32 \times 11.1 - 312 = 21 - 52 \times 11.1$; that is,

$$-556.2 = -556.2.$$

2. Given $x + 18 = 3x - 5$, to find $x$.

Transposing and reducing,

$$-2x = -23;$$

dividing both members by $-2$,

$$x = 11\tfrac{1}{2}.$$

3. Given $6 - 2x + 10 = 20 - 3x - 2$, to find $x$.

Transposing and reducing,

$$x = 2.$$

4. Given $x + \dfrac{1}{2}x + \dfrac{1}{3}x = 11$, to find $x$.

Multiplying both members by 6, and reducing,

$$11x = 66;$$

whence, $x = 6.$

5. Given $2x - \frac{1}{2}x + 1 = 5x - 2$, to find $x$.

Multiplying both members by 2, transposing and reducing,
$$-7x = -6;$$
whence, $$x = \frac{6}{7}.$$

6. Given $3ax + \frac{a}{2} - 3 = bx - a$, to find $x$.

Multiplying by 2, transposing and reducing,
$$6ax - 2bx = 6 - 3a;$$
factoring the first member of the equation, we have
$$(6a - 2b)x = 6 - 3a;$$
whence, $$x = \frac{6 - 3a}{6a - 2b}.$$

7. Given $\frac{x-3}{2} + \frac{x}{3} = 20 - \frac{x-19}{2}$, to find $x$.

Multiplying both members by 6,
$$3x - 9 + 2x = 120 - 3x + 57;$$
transposing and reducing,
$$8x = 186; \quad \therefore \quad x = 23\tfrac{1}{4}.$$

8. Given $\frac{x+3}{2} + \frac{x}{3} = 4 - \frac{x-5}{4}$, to find $x$.

Multiplying both members by 12,
$$6x + 18 + 4x = 48 - 3x + 15;$$
transposing and reducing,
$$13x = 45; \quad \therefore \quad x = 3\tfrac{6}{13}.$$

## EQUATIONS OF THE FIRST DEGREE.

9. Given $\dfrac{ax-b}{4} + \dfrac{a}{3} = \dfrac{5x}{2} - \dfrac{bx-a}{3}$, to find $x$.

Multiplying both members by 12,

$$3ax - 3b + 4a = 6bx - 4bx + 4a;$$

transposing, reducing and factoring,

$$(3a - 2b)x = 3b, \quad \therefore \quad x = \dfrac{3b}{3a - 2b}.$$

10. Given $\dfrac{3ax}{c} - \dfrac{2bx}{d} - 4 = f$, to find $x$.

Multiplying both members by $cd$,

$$3adx - 2bcx - 4cd = fcd;$$

transposing, reducing and factoring,

$$(3ad - 2bc)x = cdf + 4cd; \quad \therefore \quad x = \dfrac{cdf + 4cd}{3ad - 2bc}.$$

11. Given $\dfrac{8ax - b}{7} - \dfrac{3b - c}{2} = 4 - b$, to find $x$.

Multiplying both members by 14,

$$16ax - 2b - 21b + 7c = 56 - 14b;$$

transposing and reducing,

$$16ax = 56 + 9b - 7c; \quad \therefore \quad x = \dfrac{56 + 9b - 7c}{16a}$$

12. Given $\dfrac{x}{5} - \dfrac{x-2}{3} + \dfrac{x}{2} = \dfrac{13}{3}$, to find $x$.

Multiplying both members by 30,

$$6x - 10x + 20 + 15x = 130;$$

transposing and reducing,

$$11x = 110; \quad \therefore \quad x = 10.$$

13. Given $\dfrac{x}{a} - \dfrac{x}{b} + \dfrac{x}{c} - \dfrac{x}{d} = f$, to find $x$.

Multiplying both members by $abcd$, and factoring,

$(bcd - acd + abd - abc)x = abcdf \therefore x = \dfrac{abcdf}{bcd - acd + abd - abc}$.

14. Given $x - \dfrac{8x-5}{13} + \dfrac{4x-2}{11} = x+1$, to find $x$.

Multiplying both members by 143,

$$143x - 33x + 55 + 52x - 26 = 143x + 143;$$

transposing and reducing,

$$19x = 114; \qquad \therefore \qquad x = 6.$$

15. Given $\dfrac{x}{7} - \dfrac{8x}{9} - \dfrac{x-3}{5} = -12\dfrac{29}{45}$, to find $x$.

Multiplying both members by 315,

$$45x - 280x - 63x + 189 = -3983;$$

transposing and reducing,

$$-298x = -4172; \qquad \therefore \qquad x = 14.$$

16. Given $2x - \dfrac{4x-2}{5} = \dfrac{3x-1}{2}$, to find $x$.

Multiplying both members by 10,

$$20x - 8x + 4 = 15x - 5;$$

transposing and reducing,

$$-3x = -9; \qquad \therefore \qquad x = 3.$$

17. Given $3x + \dfrac{bx - d}{3} = x + a$, to find $x$.

Multiplying both members by 3,

$$9x + bx - d = 3x + 3a;$$

transposing, reducing and factoring,

$$(6 + b) x = 3a + d; \qquad \therefore \qquad x = \dfrac{3a + d}{6 + b}.$$

18. Given $\dfrac{(a + b)(x - b)}{a - b} - 3a = \dfrac{4ab - b^2}{a + b} - 2x + \dfrac{a^2 - bx}{b};$

to find $x$.

We see that the least common multiple of the several fractions of the two members of this equation is,

$$a^2b - b^3.$$

Hence, multiplying both members of the equation by $a^2b - b^3$, and performing all the indicated operations, we shall have,

$$a^2bx + 2ab^2x + b^3x - a^2b^2 - 2ab^3 - b^4 - 3a^3b + 3ab^3 = 4a^2b^2$$
$$- 5ab^3 + b^4 - 2a^2bx + 2b^3x + a^4 - a^2bx - a^2b^2 + b^3x;$$

then, by transposing,

$$a^2bx + 2ab^2x + b^3x + 2a^2bx - 2b^3x + a^2bx - b^3x = 4a^2b^2 - 5ab^3$$
$$+ b^4 + a^4 - a^2b^2 + a^2b^2 + 2ab^3 + b^4 + 3a^3b + 3ab^3;$$

factoring, we have,

$$2b(2a^2 + ab - b^2) x = a^4 + 3a^3b + 4a^2b^2 - 6ab^3 + 2b^4;$$

dividing by the coefficient of $x$,

$$x = \dfrac{a^4 + 3a^3b + 4a^2b^2 - 6ab^3 + 2b^4}{2b(2a^2 + ab - b^2)}.$$

19. Given, $x = 3x - \frac{1}{2}(4-x) + \frac{1}{3}.$

Clearing of fractions, and dropping parenthesis,
$$6x = 18x - 12 + 3x + 2.$$
Transposing and reducing,
$$-15x = -10;$$
$$\therefore x = \frac{2}{3}.$$

20. Given, $\dfrac{3x-7}{5} + \dfrac{25-4x}{9} = \dfrac{5x-14}{3}.$

Clearing of fractions,
$$27x - 63 + 125 - 20x = 75x - 210.$$
Transposing and reducing,
$$-68x = -272;$$
$$\therefore x = 4.$$

21. Given, $\dfrac{2x+5}{13} + \dfrac{40-x}{8} = \dfrac{10x-427}{19}.$

Clearing of fractions,
$$304x + 760 + 9880 - 247x = 1040x - 44408.$$
Transposing and reducing,
$$-983x = -55048;$$
$$\therefore x = 56.$$

22. Given, $\dfrac{x}{7} - \dfrac{x-5}{11} + 5 = x - \left(\dfrac{2x}{77} + 1\right).$

Clearing of fractions,
$$11x - 7x + 35 + 385 = 77x - 2x - 77.$$
Transposing and reducing,
$$-71x = -497;$$
$$\therefore x = 7.$$

23. Given, $\dfrac{x-1}{2} + \dfrac{x-2}{3} = \dfrac{x+3}{4} + \dfrac{x+4}{6} + 1.$

Clearing of fractions,
$$6x - 6 + 4x - 8 = 3x + 9 + 2x + 8 + 12.$$

Transposing and reducing,
$$5x = 43;$$
$$\therefore \quad x = 8\tfrac{3}{5}.$$

24. Given, $\dfrac{x-1}{x-2} - \dfrac{x-2}{x-3} = \dfrac{x-5}{x-6} - \dfrac{x-6}{x-7}.$

Performing the indicated subtraction in both members,
$$\dfrac{-1}{(x-2)(x-3)} = \dfrac{-1}{(x-6)(x-7)}.$$

Clearing of fractions, and performing indicated operations,
$$x^2 - 13x + 42 = x^2 - 5x + 6.$$

Transposing and reducing,
$$-8x = -36;$$
$$\therefore \quad x = 4\tfrac{1}{2}.$$

25. Given, $\left(x + \dfrac{5}{2}\right)\left(x - \dfrac{3}{2}\right) - (x+5)(x-3) + \dfrac{3}{4} = 0.$

Performing indicated operations,
$$x^2 + x - \dfrac{15}{4} - x^2 - 2x + 15 + \dfrac{3}{4} = 0.$$

Transposing and reducing,
$$-x = -12; \quad \therefore \quad x = 12.$$

26. Given, $\dfrac{6x+7}{15} - \dfrac{2x-2}{7x-6} = \dfrac{2x+1}{5}.$

Clearing of fractions,
$$42x^2 + 13x - 42 - 30x + 30 = 42x^2 - 15x - 18.$$

Transposing and reducing,
$$-2x = -6;$$
$$\therefore \quad x = 3.$$

27. Given, $\dfrac{1}{x-2} - \dfrac{1}{x-4} = \dfrac{1}{x-6} - \dfrac{1}{x-8}.$

Performing the indicated subtractions,
$$\dfrac{-2}{(x-2)(x-4)} = \dfrac{-2}{(x-6)(x-8)}.$$

Dividing by $-2$, and clearing of fractions,
$$x^2 - 14x + 48 = x^2 - 6x + 8.$$
Transposing and reducing,
$$-8x = -40;$$
$$\therefore \quad x = 5.$$

28. Given, $\quad (x+1)^2 = (5+x)x - 2.$
Performing indicated operations,
$$x^2 + 2x + 1 = 5x + x^2 - 2.$$
Transposing and reducing,
$$-3x = -3;$$
$$\therefore \quad x = 1.$$

29. Given, $\quad \dfrac{2}{2x-5} + \dfrac{1}{x-3} = \dfrac{6}{3x-1}.$
Clearing of fractions,
$$6x^2 - 20x + 6 + 6x^2 - 17x + 5 = 12x^2 - 66x + 90.$$
Transposing and reducing,
$$29x = 79;$$
$$\therefore \quad x = \dfrac{79}{29}.$$

30. Given, $\quad \dfrac{x}{a} + \dfrac{x}{b-a} = \dfrac{a}{b+a}.$
Factoring,
$$x\left\{\dfrac{1}{a} + \dfrac{1}{b-a}\right\} = \dfrac{a}{b+a}.$$
Reducing,
$$x\left(\dfrac{b}{a(b-a)}\right) = \dfrac{a}{(b+a)}.$$
$$\therefore \quad x = \dfrac{a^2(b-a)}{b(b+a)}.$$

31. Given, $\quad \dfrac{1}{2}\left(x - \dfrac{a}{3}\right) - \dfrac{1}{3}\left(x - \dfrac{a}{4}\right) + \dfrac{1}{4}\left(x - \dfrac{a}{5}\right) = 0.$

Performing indicated operations,
$$\frac{1}{2}x - \frac{1}{6}a - \frac{1}{3}x + \frac{1}{12}a + \frac{1}{4}x - \frac{1}{20}a = 0.$$
Clearing of fractions, and reducing,
$$25x = 8a;$$
$$\therefore \quad x = \frac{8a}{25}.$$

32. Given, $\quad 1.2x - \dfrac{.18x - .05}{.5} = .4x + 8.9.$

Clearing of fractions,
$$.6x - .18x + .05 = .2x + 4.45.$$
Transposing, and reducing,
$$.22x = 4.40;$$
$$\therefore \quad x = 20.$$

33. Given, $\quad 4.8x - \dfrac{.72x - .05}{.5} = 1.6x + 8.9.$

Clearing of fractions,
$$2.4x - .72x + .05 = 0.8x + 4.45.$$
Transposing and reducing,
$$.88x = 4.40;$$
$$\therefore \quad x = 5.$$

STATEMENT AND SOLUTION OF PROBLEMS.

8. Divide $1000 between A, B, and C, so that A shall have $72 more than B, and C $100 more than A.

Let $x$ denote the number of dollars in B's share.
Then will $x + 72$ " " " " A's "
and $x + 72 + 100$ " " " " C's "

From the conditions of the problem,

$x + x + 72 + x + 172 = 1000$; or, $3x = 756$, $\therefore x = 252$, or,   A's share is $324, B's share $252 and C's share $424.

9. A and B play together at cards. A sits down with $84 and B with $48. Each loses and wins in turn, when it appears that A has five times as much as B. How much did A win?

Let $x$ denote the number of dollars that A wins.

Then will $84 + x$ denote what A has at last,

and   $48 - x$ what B has at last;

from the conditions of the problem,

$$84 + x = 5(48 - x); \quad \text{or,} \quad 84 + x = 240 - 5x;$$

whence,   $x = 26$;   or,   A wins $26.

10. A person dying, leaves half of his property to his wife, one sixth to each of two daughters, one twelfth to a servant, and the remaining $600 to the poor: what was the amount of his property?

Let   $x$ denote the whole number of dollars in the property.

Then will $\dfrac{x}{2}$ " " " . " " in the wife's share.

$\dfrac{x}{6}$ " " " " " each daughter's "

and   $\dfrac{x}{12}$ " " " " " the servant's "

from the conditions of the problem,

$$\frac{x}{2} + 2\frac{x}{6} + \frac{x}{12} + 600 = x;$$

multiplying both members by 12, transposing and reducing,

$$-x = -7200; \quad \text{or,} \quad x = 7200.$$

11. A father leaves his property, amounting to $2520, to four sons, A, B, C and D. C is to have $360, B as much as C and D together, and A twice as much as B less $1000: how much do A, B and D receive?

Let $x$ denote the number of dollars that D receives·
Then will $x + 360$ " " " " B "
and $2x + 720 - 1000$ " " " A "

from the conditions of the problem,

$$360 + x + x + 360 + 2x + 720 - 1000 = 2520;$$

transposing and reducing,

$$4x = 2080; \qquad \therefore \qquad x = 520$$

or, D's share is $520; B's share $880, and A's share $760.

12. An estate of $7500 is to be divided between a widow, two sons, and three daughters, so that each son shall receive twice as much as each daughter, and the widow herself $500 more than all the children: what was her share, and what the share of each child?

Let $x$ denote the number of dollars in each daughter's share;·
Then will $2x$ " " " " son's "
and $4x + 3x + 500$ " " " the widow's "

from the conditions of the problem,

$$4x + 3x + 4x + 3x + 500 = 7500;$$

transposing and reducing,

$$14x = 7000; \qquad \therefore \qquad x = 500.$$

Daughters' share $500; son's share $1000; widow's share $4000.

13. A company of 180 persons consists of men, women and

children. The men are 8 more in number than the women, and the children 20 more than the men and women together: how many of each sort in the company?

Let $x$ denote the number of women;
Then will $x + 8$ "   "   men;
and $x + x + 8 + 20$ " children.

From the conditions of the problem,

$$x + x + 8 + x + x + 8 + 20 = 180;$$

transposing and reducing,

$$4x = 144; \quad \therefore \quad x = 36.$$

36 women, 44 men and 100 children.

14. A father divides $2000 among five sons, so that each elder should receive $40 more than his next younger brother: what is the share of the youngest?

Let $x$ denote the number of dollars in the youngest's share.
Then will $x + 40$ "   "   "   "   second's   "
$x + 80$ "   "   "   "   third's   "
$x + 120$ "   "   "   "   fourth's   "
$x + 160$ "   "   "   "   fifth's   "

From the conditions of the problem,

$$5x + 400 = 2000;$$

transposing and reducing,

$$5x = 1600; \quad \therefore \quad x = 320.$$

15. A purse of $2850 is to be divided among three persons, A, B and C; A's share is to be $\frac{6}{11}$ of B's share, and C is to have $300 more than A and B together: what s each one's share?

Let $x$ denote the number of dollars in B's share.

Then will $\frac{6x}{11}$ " " " " A's "

and $x + \frac{6x}{11} + 300$ " " C's "

From the conditions of the problem,

$$x + \frac{6x}{11} + x + \frac{6x}{11} + 300 = 2850;$$

clearing of fractions, transposing and reducing,

$$34x = 28050; \quad \therefore \quad x = 825;$$

hence, B's $825; A's $450; and C's $1575.

16. Two pedestrians start from the same point; the first steps twice as far as the second, but the second makes five steps while the first makes but one. At the end of a certain time they are 300 feet apart. Now, allowing each of the longer paces to be 3 feet, how far will each have travelled?

Let $x$ denote the number of feet travelled by the first.

Then will $\frac{x}{3}$ " " steps " " "

$\frac{5x}{3}$ " " " " " second.

and $1\frac{1}{2} \times \frac{5x}{3}$, or $\frac{15x}{6}$ " feet " " "

From the conditions of the problem,

$$\frac{15x}{6} - x = 300;$$

clearing of fractions, transposing and reducing,

$$9x = 1800; \quad \therefore \quad x = 200 \text{ and } \frac{15x}{6} = 500.$$

**17.** Two carpenters, 24 journeymen, and 8 apprentices, received at the end of a certain time $144. The carpenters received $1 per day, each journeyman half a dollar, and each apprentice 25 cents: how many days were they employed?

Let $x$ denote the number of days.

Then will $x$ " " dollars due each carpenter.

$\dfrac{x}{2}$ " " " " journeyman.

and $\dfrac{x}{4}$ " " " ' apprentices;

from the conditions of the problem,

$$2x + \dfrac{24x}{2} + \dfrac{8x}{4} = 144;$$

reducing,

$$16x = 144; \quad \therefore \quad x = 9.$$

**18.** A capitalist receives a yearly income of $2940; four fifths of his money bears an interest of 4 per cent., and the remainder of 5 per cent.: how much has he at interest?

Let $x$ denote the number of dollars at interest.

Then will $\dfrac{4x}{5} \times \dfrac{4}{100}$ denote the interest of 1st parcel.

and $\dfrac{x}{5} \times \dfrac{5}{100}$ " " 2d "

From the conditions of the problem,

$$\dfrac{4x}{5} \times \dfrac{4}{100} + \dfrac{x}{5} \times \dfrac{5}{100} = 2940;$$

clearing of fractions, and reducing,

$$21x = 1470000; \quad \therefore \quad x = 70000.$$

## EQUATIONS OF THE FIRST DEGREE.

19. A cistern containing 60 gallons of water has three unequal cocks for discharging it; the largest will empty it in one hour, the second in two hours, and the third in three: in what time will the cistern be emptied if they all run together?

Let $x$ denote the required number of minutes.

Then since the first emits 1 gallon per minute, the second $\frac{1}{2}$ of a gallon per minute, and the third $\frac{1}{3}$ of a gallon,

$x$ will denote the number of gallons emitted by the 1st.

$\frac{x}{2}$ " " " " " 2d.

$\frac{x}{3}$ " " " " " 3d.

From the conditions of the problem,

$$x + \frac{x}{2} + \frac{x}{3} = 60;$$

clearing of fractions and reducing,

$$11x = 360 \quad \therefore \quad x = 32\tfrac{8}{11}m.$$

20. In a certain orchard $\frac{1}{2}$ are apple-trees, $\frac{1}{4}$ peach-trees, $\frac{1}{6}$ plum-trees, 120 cherry-trees, and 80 pear-trees: how many trees in the orchard?

Let $x$ denote the whole number of trees.

Then will $\frac{x}{2}$ " " " " apple-trees.

$\frac{x}{4}$ " " . " " peach-trees.

$\frac{x}{6}$ " " " " plum-trees.

From the conditions of the problem,

$$\frac{x}{2} + \frac{x}{4} + \frac{x}{6} + 120 + 80 = x;$$

clearing of fractions, transposing and reducing,

$$-x = -2400 \quad \therefore \quad x = 2400.$$

21. A farmer being asked how many sheep he had, answered that he had them in five fields; in the 1st he had $\frac{1}{4}$, in the 2d $\frac{1}{6}$, in the 3d $\frac{1}{8}$, in the 4th $\frac{1}{12}$, and in the 5th 450: how many had he?

Let    $x$ denote the whole number of sheep:

Then will $\dfrac{x}{4}$ " " " " in 1st field.

$\dfrac{x}{6}$ " " " " 2d "

$\dfrac{x}{8}$ " " " " 3d "

and $\dfrac{x}{12}$ " " " " 4th "

From the conditions of the problem,

$$\frac{x}{4} + \frac{x}{6} + \frac{x}{8} + \frac{x}{12} + 450 = x;$$

multiplying both members by 24, transposing and reducing,

$$-9x = -10800 \quad \therefore \quad x = 1200.$$

22. My horse and saddle together are worth \$132, and the horse is worth ten times as much as the saddle: what is the value of the horse?

Let    $x$ denote the number of dollars that the saddle is worth.

Then will $10x$ " " " " horse "

From the conditions of the problem,

$$x + 10x = 132;$$

reducing, $11x = 132$ $\therefore$ $x = 12$; whence, $10x = 120$.

23. The rent of an estate is this year 8 per cent. greater than it was last. This year it is $1890: what was it last year?

Let $x$ denote the number of dollars in last year's rent.

Then will $x + \dfrac{8x}{100}$ " " " this " "

From the conditions of the problem,

$$x + \frac{8x}{100} = 1890;$$

clearing of fractions and reducing,

$$108x = 189000; \quad \therefore \quad x = 1750.$$

24. What number is that from which, if 5 be subtracted, $\frac{2}{3}$ of the remainder will be 40?

Let $x$ denote the number required:

From the conditions of the problem,

$$\tfrac{2}{3}(x - 5) = 40;$$

clearing of fractions, performing operations indicated, transposing and reducing,

$$2x = 130; \quad \therefore \quad x = 65.$$

25. A post is $\frac{1}{4}$ in the mud, $\frac{1}{3}$ in the water, and ten feet above the water: what is the whole length of the post?

Let $x$ denote the number of feet in length.

Then will $\dfrac{x}{4}$ " " " " the mud;

and $\dfrac{x}{3}$ " " " " the water:

From the conditions of the problem;

$$\frac{x}{4} + \frac{x}{3} + 10 = x;$$

clearing of fractions, transposing and reducing,

$$-5x = -120; \quad \therefore \quad x = 24.$$

26. After paying $\frac{1}{4}$ and $\frac{1}{5}$ of my money, I had 66 guineas left in my purse: how many guineas were in it at first?

Let $x$ denote the number at first;

from the conditions of the problem,

$$x - \frac{x}{4} - \frac{x}{5} = 66;$$

clearing of fractions, transposing and reducing,

$$11x = 1320; \quad \therefore \quad x = 120.$$

27. A person was desirous of giving 3 pence apiece to some beggars, but found he had not money enough in his pocket by 8 pence; he therefore gave them each two pence and had 3 pence remaining: required the number of beggars.

Let $x$ denote the number of beggars;
then, by the first condition,

$$3x - 8 \quad \text{denotes the number of pence;}$$

by the second condition,

$$2x + 3 \quad \text{denotes the number of pence;}$$

hence, $\quad 3x - 8 = 2x + 3;$

transposing and reducing,

$$x = 11.$$

## EQUATIONS OF THE FIRST DEGREE.

28. A person in play lost $\frac{1}{4}$ of his money, and then won 3 shillings; after which he lost $\frac{1}{3}$ of what he then had; and this done, found that he had but 12 shillings remaining: what had he at first?

Let $x$ denote the number of shillings at first;

Then will $x - \frac{x}{4} + 3$ " " " after first sitting;

and $\left(x - \frac{x}{4} + 3\right) - \frac{1}{3}\left(x - \frac{x}{4} + 3\right)$

will denote what he finally had;

hence, from the conditions of the problem,

$$x - \frac{x}{4} + 3 - \frac{1}{3}\left(x - \frac{x}{4} + 3\right) = 12;$$

clearing of fractions, performing indicated operations, transposing and reducing,

$$6x = 120; \quad \therefore \quad x = 20.$$

29. Two persons, A and B, lay out equal sums of money in trade; A gains $126, and B loses $87, and A's money is now double B's: what did each lay out?

Let $x$ denote the number of dollars laid out by each;
Then will $x + 126$ " " " A had;
and $x - 87$ " " " B "

From the conditions of the problem,

$$x + 126 = 2(x - 87);$$

performing indicated operations, transposing and reducing,

$$-x = -300; \quad \therefore \quad x = 300.$$

30. A person goes to a tavern with a certain sum of money in his

pocket, where he spends 2 shillings; he then borrows as much money as he had left, and going to another tavern, he there spends 2 shillings also; then borrowing again as much money as was left, he went to a third tavern, where, likewise, he spent 2 shillings and borrowed as much as he had left; and again spending 2 shillings at a fourth tavern, he then had nothing remaining. What had he at first?

Let $x$ denote the number of shillings at first.
Then, from the first condition,

$x - 2$ will denote what he has after 1st visit.
$2(x - 2) - 2$ or $2x - 6$ " " " " 2d "
$2(2x - 6) - 2$ or $4x - 14$ " " " " 3d "
$2(4x - 14) - 2$ or $8x - 30$ " " " " 4th "

From the conditions of the problem,

$8x - 30 = 0$ or $8x = 30$; $\therefore x = 3\frac{3}{4}$.

or the amount at first was $3s.\ 9d.$

31. A farmer bought a basket of eggs, and offered them at 7 cents a dozen. But before he had sold any, 5 dozen were broken by a careless boy, for which he was paid. He then sold the remainder at 8 cents a dozen, and received as much as he would have got for the whole at the first price. How many eggs had he in his basket?

Let $x$ denote the number of dozens at first;
Then will $x - 5$ " " " sold;
and $8(x - 5)$ " " cents received;
$7x$ " " " first asked;
hence, $7x = 8(x - 5)$; $\therefore x = 40$.

32. A cask, A, contains a mixture of 12 gallons of wine and 18 gallons of water; another cask, B, contains a mixture of 9 gallons of wine and 3 gallons of water: how many gallons must be drawn from each to produce a mixture of 7 gallons of wine and 7 gallons of water?

Let $x$ denote the number of gallons drawn from the cask A, and $14-x$ the number of gallons drawn from the cask B.

Of the $x$ gallons drawn from A, $\frac{12}{30}$ths is wine and $\frac{18}{30}$ths is water; in like manner, of the mixture drawn from B, $\frac{9}{12}$ths is wine, and $\frac{3}{12}$ths is water. Hence, all the wine drawn from both is equal to $\frac{12}{30}x + \frac{9}{12}(14-x)$ gallons; but this is equal to 7 gallons. Hence,

$$\frac{12}{30}x + \frac{9}{12}(14-x) = 7,$$

or, 
$$\frac{2}{5}x + \frac{3}{4}(14-x) = 7;$$

∴ $x = 10$ and $14 - x = 4$.

33. At what time between 1 and 2 o'clock is the minute hand of a clock just 1 minute space ahead of the hour hand?

Let $x$ denote the number of minute spaces passed over by the hour hand from 12 o'clock till the hands have the required position. Then $61 + x$ will denote the number of minute spaces passed over by the minute hand in the same time; but the minute hand travels 12 times as fast as the hour hand. Hence,

$$12x = 61 + x,$$
or, $$11x = 61;$$

∴ $x = 5\frac{6}{11}$ and $61 + x = 66\frac{6}{11}$.

That is, the hands will have the required position at 1 h. $6\frac{6}{11}$ min.

34. A person having $a$ hours at his disposal, how far can he ride in a coach that travels $b$ miles per hour, and return on foot at the rate of $c$ miles per hour?

Let $x$ denote the number of miles.

The time required to ride $x$ miles in the coach will be denoted by $\dfrac{x}{b}$, and the time required to walk back will be denoted by $\dfrac{x}{c}$. From the conditions of the problem, we have,

$$\frac{x}{b} + \frac{x}{c} = a. \text{ Hence,}$$

$$\left[(b+c)x = abc; \quad \therefore \quad x = \frac{abc}{b+c}.\right]$$

35. A can do a piece of work in one-half the time that B can; and B can do it in two-thirds the time that C can; all together can do it in 6 days. How many days would it take each to do it singly?

Let $x$ denote the number that it will take A to do it. Then will $2x$ denote the number of days that it will take B to do it; and $3x$ will denote the number of days it will take C to do it. Consequently, A can do a part denoted by $\dfrac{1}{x}$ in one day, B can do a part denoted by $\dfrac{1}{2x}$, C can do a part denoted by $\dfrac{1}{3x}$ in the same time, and all together can do a part denoted by $\dfrac{1}{6}$ in one day. Hence, from the conditions of the problem,

$$\frac{1}{x} + \frac{1}{2x} + \frac{1}{3x} = \frac{1}{6}. \text{ Factoring,}$$

$$\frac{1}{x}\left(1 + \frac{1}{2} + \frac{1}{3}\right) = \frac{1}{6}, \text{ or, } \frac{1}{x}(6 + 3 + 2) = 1;$$

$$\therefore \ \frac{1}{x} = \frac{1}{11}, \text{ or, } x = 11, \ 2x = 22, \text{ and } 3x = 33.$$

## SIMULTANEOUS EQUATIONS OF THE FIRST DEGREE.

1. Given $\begin{cases} 2x + 3y = 16 \\ 3x - 2y = 11 \end{cases}$ to find $x$ and $y$.

Multiply both members of the first by 2, and of the second by 3;

$$\begin{matrix} 4x + 6y = 32 \\ 9x - 6y = 33 \end{matrix}$$

whence, by addition, member to member, we have,

$13x = 65;$ $\therefore$ $x = 5,$ also, $y = 2.$

2. Given $\begin{cases} \dfrac{2x}{5} + \dfrac{3y}{4} = \dfrac{9}{20} \\ \dfrac{3x}{4} + \dfrac{2y}{5} = \dfrac{61}{120} \end{cases}$ to find $x$ and $y$.

Clearing of fractions, and then multiplying both members of the first by 16, and of the second by 5,

$$\begin{matrix} 128x + 240y = 144 \\ 450x + 240y = 305 \end{matrix}$$

whence, by subtracting, member from member,

$322x = 161;$ $\therefore$ $x = \dfrac{1}{2},$ also, by substitution, $y = \dfrac{1}{3}.$

3. Given $\begin{cases} \dfrac{x}{7} + 7y = 99 \\ \dfrac{y}{7} + 7x = 51 \end{cases}$ to find $x$ and $y$.

Multiplying the first by 343 and the second by 7;

$$\begin{matrix} 49x + 2401y = 33957 \\ y + 49x = 357 \end{matrix}$$

by subtraction,

$2400y = 33600$; ∴ $y = 14$; also, by substitution, $x = 7$.

4. Given $\begin{cases} \dfrac{x}{2} - 12 = \dfrac{y}{4} + 8 \\ \dfrac{x+y}{5} + \dfrac{x}{3} - 8 = \dfrac{2y-x}{4} + 27 \end{cases}$; to find $x$ and $y$.

Clearing of fractions and transposing,

$$2x - y = 80$$
$$47x - 18y = 2100;$$

multiplying both members of the first by 18, and subtracting the result from the second, member from member,

$11x = 660$; ∴ $x = 60$; by substitution, $y = 40$.

5. Given $\begin{cases} x + y + z = 29 & \cdots & (1) \\ x + 2y + 3z = 62 & \cdots & (2) \\ \dfrac{x}{2} + \dfrac{y}{3} + \dfrac{z}{4} = 10 & \cdots & (3) \end{cases}$ to find $x$, $y$ and $z$.

Combining (1) and (2),
$$y + 2z = 33 \cdots (4);$$
combining (1) and (3)
$$2y + 3z = 54 \cdots (5);$$
combining (4) and (5)

$z = 12$; by successive substitutions, $x = 8$, $y = 9$.

6. Given $\begin{cases} 2x + 4y - 3z = 22 & \cdots & (1) \\ 4x - 2y + 5z = 18 & \cdots & (2) \\ 6x + 7y - z = 63 & (3) \end{cases}$; to find $x$, $y$, and $z$.

Combining (1) and (2),

$$10y - 11z = 26 \quad . \quad . \quad . \quad (4);$$

combining (1) and (3),

$$5y - 8z = 3 \quad . \quad . \quad . \quad (5);$$

combining (4) and (5),

$$5z = 20; \quad \therefore \quad z = 4.$$

By successive substitutions, $\quad x = 3, \quad y = 7.$

7. Given
$$\left\{ \begin{array}{l} x + \dfrac{y}{2} + \dfrac{z}{3} = 32 \\[4pt] \dfrac{x}{3} + \dfrac{y}{4} + \dfrac{z}{5} = 15 \\[4pt] \dfrac{x}{4} + \dfrac{y}{5} + \dfrac{z}{6} = 12 \end{array} \right\} ; \text{ to find } x, y \text{ and } z.$$

Clearing of fractions,

$$6x + 3y + 2z = 192 \quad . \quad . \quad . \quad (1);$$
$$20x + 15y + 12z = 900 \quad . \quad . \quad . \quad (2);$$
$$15x + 12y + 10z = 720 \quad . \quad . \quad . \quad (3);$$

combining (1) and (2),

$$16x + 3y = 252 \quad . \quad . \quad . \quad (4);$$

combining (1) and (3),

$$15x + 3y = 240 \quad . \quad . \quad . \quad (5);$$

combining (4) and (5),

$$x = 12;$$

by successive substitutions, $y = 20, \quad z = 30.$

8. Given
$$\left\{ \begin{array}{l} 7x - 2z + 3u = 17 \quad . \quad . \quad . \quad (1) \\ 4y - 2z + t = 11 \quad . \quad . \quad . \quad (2) \\ 5y - 3x - 2u = 8 \quad . \quad . \quad . \quad (3) \\ 4y - 3u + 2t = 9 \quad . \quad . \quad . \quad (4) \\ 3z + 8u = 33 \quad . \quad . \quad . \quad (5) \end{array} \right\} ; \text{ to find } x, y, z, t \text{ and } u.$$

Combining (2) and (4),
$$4y - 4z + 3u = 13 \quad \ldots \quad (6);$$
combining (1) and (3),
$$35y - 6z - 5u = 107 \quad \ldots \quad (7);$$
combining (5) and (6),
$$12y + 41u = 171 \quad \ldots \quad (8);$$
combining (5) and (7),
$$35y + 11u = 173 \quad \ldots \quad (9);$$
combining (8) and (9),
$$1303u = 3909; \quad \therefore \quad u = 3;$$
by successive substitutions, $x = 2$, $y = 4$, $z = 3$, $t = 1$.

9. Given,
$$\begin{cases} 3x + 2y - 4z = 15 & \ldots \ldots (1) \\ 5x - 3y + 2z = 28 & \ldots \ldots (2) \\ 3y + 4z - x = 24 & \ldots \ldots (3) \end{cases}$$

Combining (1) and (3),
$$11y + 8z = 87 \quad \ldots \ldots (4)$$
Combining (2) and (3),
$$12y + 22z = 148 \quad \ldots \ldots (5)$$
Combining (4) and (5),
$$73y = 365; \quad \therefore \quad y = 5.$$
Substituting and reducing,
$$x = 7, \text{ and } z = 4.$$

10. Given,
$$\begin{cases} \dfrac{1}{x} + \dfrac{1}{y} = 1 & \ldots \ldots (1) \\ \dfrac{1}{x} + \dfrac{1}{z} = 2 & \ldots \ldots (2) \\ \dfrac{1}{y} + \dfrac{1}{z} = \dfrac{3}{2} & \ldots \ldots (3) \end{cases}$$

Adding (1), (2), and (3), member to member, and dividing by 2,
$$\frac{1}{x} + \frac{1}{y} + \frac{1}{z} = \frac{9}{4} \quad \ldots \ldots \ldots (4)$$

Subtracting (1), (2), and (3), successively, from (4),
$$\frac{1}{z} = \frac{5}{4}; \quad \therefore \; z = \frac{4}{5}.$$
$$\frac{1}{y} = \frac{1}{4}; \quad \therefore \; y = 4.$$
$$\frac{1}{x} = \frac{3}{4}; \quad \therefore \; x = \frac{4}{3}.$$

11. Given,
$$\begin{cases} \dfrac{2}{x} + \dfrac{1}{y} = \dfrac{3}{z} & \ldots \ldots \ldots (1) \\ \dfrac{3}{z} - \dfrac{2}{y} = 2 & \ldots \ldots \ldots (2) \\ \dfrac{1}{x} + \dfrac{1}{2} = \dfrac{4}{3} & \ldots \ldots \ldots (3) \end{cases}$$

Making $\dfrac{1}{x} = x'$, $\dfrac{1}{y} = y'$, and $\dfrac{1}{z} = z'$,
$$2x' + y' = 3z' \quad \ldots \ldots \ldots (4)$$
$$3z' - 2y' = 2 \quad \ldots \ldots \ldots (5)$$
$$x' + z' = \frac{4}{3} \quad \ldots \ldots \ldots (6)$$

Combining (4) and (5),
$$4x' + 3z' = 6z' + 2,$$
or,
$$4x' - 3z' = 2 \quad \ldots \ldots \ldots (7)$$

Combining (6) and (7),
$$7x' = 6; \quad \therefore \; x' = \frac{6}{7}, \quad \text{or,} \quad x = \frac{7}{6}.$$

Substituting and reducing,
$$z' = \frac{10}{21}, \quad \text{or,} \quad z = \frac{21}{10},$$
and,
$$y' = -\frac{2}{7}, \quad \text{or,} \quad y = -\frac{7}{2}.$$

12. Given,
$$\begin{cases} \dfrac{3y-1}{4} = \dfrac{6z}{5} - \dfrac{x}{2} + \dfrac{9}{5} & \dots \dots (1) \\ \dfrac{5x}{4} + \dfrac{4z}{3} = y + \dfrac{5}{6} & \dots \dots (2) \\ \dfrac{3x+1}{7} - \dfrac{z}{14} + \dfrac{1}{6} = \dfrac{2z}{21} + \dfrac{y}{3} & \dots (3) \end{cases}$$

Clearing of fractions and transposing,

$$15y + 10x - 24z = 41 \dots \dots (4)$$
$$-12y + 15x + 16z = 10 \dots \dots (5)$$
$$-14y + 18x - 7z = -13 \dots \dots (6)$$

Combining (4) and (5),
$$115x - 16z = 214 \dots \dots (7)$$

Combining (4) and (6),
$$410x - 441z = 379 \dots \dots (8)$$

Combining (7) and (8),
$$8831z = 8831;$$
$$\therefore z = 1.$$

Substituting and reducing,
$$x = 2, \text{ and } y = 3.$$

13. Given,
$$\begin{cases} \dfrac{x}{a} + \dfrac{y}{b} = 1 & \dots \dots (1) \\ \dfrac{x}{a} + \dfrac{z}{c} = 1 & \dots \dots (2) \\ \dfrac{y}{b} + \dfrac{z}{c} = 1 & \dots \dots (3) \end{cases}$$

Adding (1), (2), and (3), and dividing by 2,
$$\dfrac{x}{a} + \dfrac{y}{b} + \dfrac{z}{c} = \dfrac{3}{2} \dots \dots (4)$$

Subtracting (1), (2), and (3), successively, from (4),

$$\frac{z}{c} = \frac{1}{2}; \quad \therefore \quad z = \frac{c}{2},$$

$$\frac{y}{b} = \frac{1}{2}; \quad \therefore \quad y = \frac{b}{2},$$

$$\frac{x}{a} = \frac{1}{2}; \quad \therefore \quad x = \frac{a}{2}.$$

14. Given,
$$\begin{cases} 7x - 3y = 1 & \ldots \ldots (1) \\ 11z - 7u = 1 & \ldots \ldots (2) \\ 4z - 7y = 1 & \ldots \ldots (3) \\ 19x - 3u = 1 & \ldots \ldots (4) \end{cases}$$

Combining (1) and (4),
$$21u - 57y = 12 \ldots \ldots (5)$$

Combining (2) and (3),
$$28u - 77y = 7 \ldots \ldots (6)$$

Combining (5) and (6),
$$3y = 27; \quad \therefore \quad y = 9.$$

Substituting and reducing,
$$x = 4, \quad z = 16, \text{ and } \quad u = 25.$$

15. Given,
$$\begin{cases} \dfrac{3x - 5y}{2} + 3 = \dfrac{2x + y}{5} & \ldots \ldots (1) \\ 8 - \dfrac{x - 2y}{4} = \dfrac{x}{2} + \dfrac{y}{3} & \ldots \ldots (2) \end{cases}$$

Clearing of fractions and transposing,
$$11x - 27y = -30 \ldots \ldots (3)$$
$$9x - 2y = 96 \ldots \ldots (4)$$

$$\therefore \quad y = 6, \text{ and } \quad x = 12.$$

16. Given,
$$\begin{cases} \dfrac{3x}{10} - \dfrac{y}{15} - \dfrac{4}{9} = \dfrac{x}{12} - \dfrac{y}{18} & \dots \ (1) \\ 2x - \dfrac{8}{3} = \dfrac{x}{12} - \dfrac{y}{15} + \dfrac{11}{10} & \dots \ (2) \end{cases}$$

Clearing of fractions, transposing, and reducing,

$$39x - 2y = 80 \ \dots \dots \ (3)$$

$$115x + 4y = 226 \ \dots \dots \ (4)$$

Combining (3) and (4),

$$193x = 386;$$

$$x = 2, \text{ and } y = -1.$$

## PROBLEMS GIVING RISE TO SIMULTANEOUS EQUATIONS OF THE FIRST DEGREE.

5. What two numbers are they, whose sum is 33 and whose difference is 7?

Let $x$ denote the first, and $y$ the second.
From the conditions,

$$x + y = 33$$
$$x - y = \ 7;$$

whence, by combination,

$$x = 20, \qquad y = 13.$$

6. Divide the number 75 into two such parts, that three times the greater may exceed seven times the less by 15.

Let $x$ denote the greater, and $y$ the less
From the conditions of the problem,

$$x + y = 75$$
$$3x - 7y = 15;$$

by combination, $10y = 210$;  $\therefore y = 21$; also, $x = 54$.

7. In a mixture of wine and cider, $\frac{1}{2}$ of the whole plus 25 gallons was wine, and $\frac{1}{3}$ part minus 5 gallons, was cider: how many gallons were there of each?

Let  $x$ denote the number of gallons of wine;
and  $y$  "    "    "    cider;
then will $x + y$  "    "    mixture.

From the conditions,

$$\frac{x+y}{2} + 25 = x$$

$$\frac{x+y}{3} - 5 = y;$$

clearing of fractions, transposing and reducing,

$$y - x = -50$$
$$-2y + x = 15$$

by combination,

$$y = 35; \quad \text{and} \quad x = 85$$

8. A bill of £120 was paid in guineas and moidores, and the number of pieces of both sorts that were used was just 100; if the guineas were estimated at 21s., and the moidores at 27s., how many were there of each?

Let  $x$ denote the number of moidores;
and  $y$  "    "    " guineas;

then, since £120 $= 2400s.$, we have, from the conditions,

$$x + y = 100$$
$$27x + 21y = 2400;$$

by combination, $6y = 300$; $\therefore y = 50$; also, $x = 50$.

9. Two travellers set out at the same time from London and York, whose distance apart is 150 miles; they travel toward each other; one of them goes 8 miles a day, and the other 7; in what time will they meet?

Let $x$ denote the number of miles travelled by the first;

$y$ " " " " " second;

then will $\dfrac{x}{8}$ " " days " " first;

and $\dfrac{y}{7}$ " " " " " second;

From the conditions,
$$x + y = 150;$$
$$\dfrac{x}{8} = \dfrac{y}{7};$$

whence, by combination,

$x = 80$ and $\dfrac{x}{8} = 10$, the number of days.

10. At a certain election, 375 persons voted for two candidates; and the candidate chosen had a majority of 91; how many voted for each?

Let $x$ denote the number of votes received by the first;

$y$ " " " " " second;

from the conditions of the problem,
$$x + y = 375$$
$$x = y + 91;$$

by combination, $x = 233$, $y = 142$.

11. A's age is double B's, and B's is triple C's, and the sum of all their ages is 140: what is the age of each?

Let $x$ denote the age of A;
$y$ " " B;
$z$ " " C;

from the conditions of the problem,

$$x = 2y \quad \cdots \quad (1);$$
$$y = 3z \quad \cdots \quad (2);$$
$$x + y + z = 140 \quad \cdots \quad (3);$$

from (1) and (2), $x = 6z$;
substituting, $y = 3z$, and $x = 6z$, in (3), and reducing,
$10z = 140$; ∴ $z = 14$, $x = 84$, $y = 42$.

12. A person bought a chaise, horse and harness, for £60; the horse came to twice the price of the harness, and the chaise to twice the price of the horse and harness: what did he give for each?

Let $x$ denote the number of pounds paid for the harness;
$y$ " " " " " horse;
$z$ " " " " " chaise;

from the conditions of the problem,

$$y = 2x \quad \cdots \quad (1)$$
$$z = 2(x + y) \quad \cdots \quad (2)$$
$$x + y + z = 60 \quad \cdots \quad (3)$$

from (2) and (1) $z = 6x$;
substituting $z = 6x$ and $y = 2x$ in (3)
$9x = 60$, ∴ $x = 6\frac{2}{3}$, also, by substitution, $y = 13\frac{1}{3}$, $z = 40$;
hence, the price of the chaise was £40; of the horse £13 6s. 8d.; and that of the harness £6 13s. 4d.

13. A person has two horses, and a saddle worth £50; now, if the saddle be put on the back of the first horse, it will make his value double that of the second; but if it be put on the back of the second, it will make his value triple that of the first: what is the value of each horse?

Let $x$ denote the number of pounds the 1st horse is worth;

$y$ "        "        '        " 2d "        "

from the conditions of the problem,

$$x + 50 = 2y$$
$$y + 50 = 3x;$$

whence, by combination,

$$x = 30 \quad y = 40.$$

14. Two persons, A and B, have each the same income. A saves ⅕ of his yearly; but B, by spending £50 per annum more than A, at the end of 4 years finds himself £100 in debt; what is the income of each?

Let $x$ denote the number of pounds in the income of A ;

$y$ "        "        "        "        " B ;

by the conditions of the problem, these are equal; one only will be used. Then will

$\frac{4}{5}x$ denote what A spends per year ;

$\frac{4}{5}x + 50$ "    " B    "        "

from the conditions of the problem,

$$4\left(\frac{4}{5}x + 50\right) = 4x + 100;$$

whence, performing indicated operations, transposing and reducing,

$$4x = 500 \quad \therefore \quad x = 125.$$

15. To divide the number 36 into three such parts, that $\frac{1}{2}$ of the first, $\frac{1}{3}$ of the second, and $\frac{1}{4}$ of the third, may be all equal to each other.

Let $x$, $y$ and $z$, denote the parts.
From the conditions of the problem,

$$x + y + z = 36$$

$$\frac{x}{2} = \frac{y}{3}$$

$$\frac{x}{2} = \frac{z}{4};$$

clearing of fractions, and combining,

$9x = 72$ ∴ $x = 8$; whence, $y = 12$ and $z = 16$.

16. A footman agreed to serve his master for £8 a year and livery, but was turned away at the end of 7 months, and received only £2 13s. 4d. and his livery: what was its value?

Let $x$ denote the value of livery, expressed in shillings: £8 = 160s., and £2 13s. 4d. = $53\frac{1}{3}$s.;

Then will $\left(\dfrac{160 + x}{12}\right)$ denote the value of wages 1 month,

and $7\left(\dfrac{160 + x}{12}\right)$ " ' " 7

by the conditions of the problem,

$$7\left(\frac{160 + x}{12}\right) = 53\frac{1}{3} + x$$

$$1120 + 7x = 640 + 12x$$

$$-5x = -480$$

$$x = 96 \quad \therefore \quad \text{value, £4.16s.}$$

17. To divide the number 90 into four such parts, that if the first be increased by 2, the second diminished by 2, the third multiplied by 2, and the fourth divided by 2, the sum, difference, product and quotient, so obtained, will be all equal to each other.

Let $x$, $y$, $z$ and $u$, denote the parts; from the conditions of the problem,

$$x + y + z + u = 90$$
$$x + 2 = y - 2$$
$$x + 2 = 2z$$
$$x + 2 = \frac{u}{2};$$

whence we find from the last three equations,

$$y = x + 4, \quad z = \frac{x}{2} + 1, \quad \text{and} \quad u = 2x + 4;$$

substituting these values in the first equation,

$$x + x + 4 + \frac{x}{2} + 1 + 2x + 4 = 90; \text{ or } 4\tfrac{1}{2}x = 81; \therefore x = 18;$$

whence, by substitution, $y = 22$, $z = 10$, and $u = 40$.

18. The hour and minute hands of a clock are exactly together at 12 o'clock: when are they next together.

### 1st Solution.

Let $x$ denote the number of minute spaces passed by the hour hand before they come together;

and $y$ the number passed by the minute hand;

then, since the latter travels 12 times as fast as the former, and since it has to gain 60 spaces, we have,

$$x - y = 60$$
$$x = 12y;$$

by combination,

$11y = 60 \quad \therefore \quad y = 5\tfrac{5}{11}; \quad \text{also,} \quad x = 65\tfrac{5}{11};$

hence, they will be together, $65\tfrac{5}{11}$ minutes after 12 o'clock, or at 1 o'clock, $\tfrac{5}{11}$ minutes, and at the end of every succeeding equal portion of time.

### 2d Solution.

The minute hand will pass the hour hand 11 times before they again come together at 12 o'clock, and the times between any two consecutive coincidences will be equal. Hence each time will be equal to 12 hours divided by $11 = 1\tfrac{1}{11} hr. = 1 hr.\ 5\tfrac{1}{11} m.$

19. A man and his wife usually drank out a cask of beer in 12 days; but when the man was from home, it lasted the woman 30 days; how many days would the man be in drinking it alone?

Let $x$ denote the number of days it takes the man to drink it;

$y$ " " " " woman " "

then, if the whole quantity of beer be denoted by 1,

$\dfrac{1}{x}$ will denote the quantity drank by the man in 1 day; and

$\dfrac{1}{y}$ " " " " woman;

from the conditions of the problem,

$$\frac{1}{x} + \frac{1}{y} = \frac{1}{12}$$

$$\frac{1}{y} = \frac{1}{30};$$

substituting the value of $\dfrac{1}{y}$ in the first equation,

$$\frac{1}{x} + \frac{1}{30} = \frac{1}{12};$$

clearing of fractions,

$$60 + 2x = 5x; \qquad \therefore \qquad x = 20.$$

**20.** *If A and B together can perform a piece of work in 8 days, A and C together in 9 days, and B and C in 10 days: how many days would it take each person to perform the same work alone?*

Let the work be denoted by 1;

Let $x$ denote the work done by A in one day;
$y$ " " " B " "
$z$ " " " C " "

then will $\dfrac{1}{x}, \dfrac{1}{y}$ and $\dfrac{1}{z}$ respectively denote the number of days that it will take A, B, and C severally to do the work;

from the conditions of the problem,

$$x + y = \tfrac{1}{8} \quad \cdots \quad (1)$$
$$x + z = \tfrac{1}{9} \quad \cdots \quad (2)$$
$$y + z = \tfrac{1}{10} \quad \cdots \quad (3);$$

clearing of fractions,

$$8x + 8y = 1 \quad \cdots \quad (4)$$
$$9x + 9z = 1 \quad \cdots \quad (5)$$
$$10y + 10z = 1 \quad \cdots \quad (6);$$

combining (4) and (5),

$$72y - 72z = 1 \quad \cdots \quad (7),$$

combining (6) and (7),

$$1440y = 82 \qquad \therefore \qquad y = \tfrac{41}{720};$$

substituting in (1) and (3),

$$x = \tfrac{1}{8} - \tfrac{41}{720} = \tfrac{49}{720}; \qquad z = \tfrac{1}{10} - \tfrac{41}{720} = \tfrac{31}{720};$$

hence, $\dfrac{1}{x} = 14\tfrac{34}{49}; \qquad \dfrac{1}{y} = 17\tfrac{23}{41}; \qquad \dfrac{1}{z} = 23\tfrac{7}{3}.$

## EQUATIONS OF THE FIRST DEGREE.

**21.** A laborer can do a certain work expressed by $a$, in a time expressed by $b$; a second laborer, the work $c$ in a time $d$; a third, the work $e$ in a time $f$. Required the time it would take the three laborers, working together, to perform the work $g$?

If a laborer can do a piece of work denoted by $a$, in a number of days denoted by $b$, he can do in 1 day so much of the work as is denoted by $\frac{a}{b}$; the second in 1 day can do so much as is denoted by $\frac{c}{d}$; and the third so much as is denoted by $\frac{e}{f}$; hence, the three working together can do

$$\frac{a}{b} + \frac{c}{d} + \frac{e}{f} = \frac{adf + bcf + bde}{bdf}$$

Let $x$ denote the time required to perform the work $g$; then, the three can perform the work $\frac{g}{x}$ in the time 1; from the conditions of the problem,

$$\frac{g}{x} = \frac{adf + bcf + bde}{bdf}$$

taking the reciprocals of each member, and then clearing of fractions, we have,

$$x = \frac{bdfg}{adf + bcf + bde}$$

In this example only a single unknown quantity has been used, and it may be remarked that many other examples, in this chapter, may be more easily solved by a single unknown quantity; in such cases more than one has been used for the purpose of illustration.

**22.** If 32 pounds of sea water contain 1 pound of salt, how much fresh water must be added to these 32 pounds, in order that the quantity of salt contained in 32 pounds of the new mixture shall be reduced to 2 ounces, or $\frac{1}{8}$ of a pound?

Let $x$ denote the number of pounds to be added; then will $\dfrac{1}{32+x}$ denote the number of pounds of salt in each pound of the mixture, but this we know to be $\dfrac{1}{32} \times \dfrac{1}{8}$, or, $\dfrac{1}{256}$; hence from the conditions of the problem,

$$\dfrac{1}{32+x} = \dfrac{1}{256}, \quad \text{or,} \quad 32 + x = 256, \quad \text{or,} \quad x = 224.$$

This problem is also solved by a single unknown quantity more readily than by two.

23. A number is expressed by three figures; the sum of these figures is 11; the figure in the place of units is double that in the place of hundreds; and when 297 is added to this number, the sum obtained is expressed by the figures of this number reversed. What is the number?

Let $x$, $y$ and $z$ denote the digits in their order; then will the number be denoted by

$$100x + 10y + z;$$

from the conditions of the problem,

$$x + y + z = 11 \quad \cdots \cdots \cdots \quad (1)$$
$$z = 2x \quad \cdots \cdots \cdots \quad (2)$$
$$100x + 10y + z + 297 = 100z + 10y + x \quad \cdots \quad (3);$$

reducing (3), gives

$$99z - 99x = 297 \quad \cdots \cdots \cdots \quad (4);$$

substituting $z = 2x$ in (4), and reducing,

$$99x = 297; \quad \therefore \quad x = 3;$$

whence, by successive substitutions,

$$y = 2, \quad z = 6. \qquad Ans.\ 326.$$

24. A person who possessed $100000 dollars, placed the greater part of it out at 5 per cent. interest, and the other part at 4 per cent. The interest which he received for the whole amounted to 4640 dollars. Required the two parts.

Let $x$ denote the greater part;

$y$ " " lesser "

From the conditions of the problem,

$$\frac{5x}{100} = \text{interest on } x \text{ dollars at 5 per cent.};$$

$$\frac{4y}{100} = \quad " \quad y \quad " \quad 4 \quad " \quad " \quad \text{then,}$$

$$x + y = 100000 \ \cdot\ \cdot\ \cdot\ \cdot\ (1)$$

$$\frac{5x}{100} + \frac{4y}{100} = 4640 \ \cdot\ \cdot\ \cdot\ \cdot\ (2)$$

clearing (2) of fractions,

$$5x + 4y = 464000 \ \cdot\ \cdot\ \cdot\ \cdot\ (3)$$

combining (1) and (3),

$$y = 36000, \qquad \text{whence} \qquad x = 64000.$$

25. A person possessed a certain capital, which he placed out at a certain interest. Another person possessed 10000 dollars more than the first, and putting out his capital 1 per cent. more advantageously, had an income greater by 800 dollars. A third, possessed 15000 dollars more than the first, and putting out his capital 2 per cent. more advantageously, had an income greater by 1500 dollars. Required the capitals and the three rates of interest.

Let $x$ denote the number of dollars in 1st capital;
and $y$ the rate per cent.; then,

$$\frac{(x \times y)}{100}$$ will denote the number of dollars of 1st income;

$$\frac{(x + 10000)(y + 1)}{100}$$ " " " " 2d "

$$\frac{(x + 15000)(y + 2)}{100}$$ " " " " 3d "

from the conditions of the problem,

$$\frac{(x + 10000)(y + 1)}{100} = \frac{x \times y}{100} + 800$$

$$\frac{(x + 15000)(y + 2)}{100} = \frac{x \times y}{100} + 1500;$$

clearing of fractions, performing indicated operations, transposing and reducing,

$$10000y + x = 70000$$
$$15000y + 2x = 120000;$$

combining and reducing,

$$5000y = 20000 \quad \therefore \quad y = 4; \quad \text{and}$$

by substitution, $x = 30000;$

2d. $40000, rate 5 per cent.
3d. $45000, rate 6 " "

26. A cistern may be filled by three pipes, A, B, C. By the two first it can be filled in 70 minutes; by the first and third it can be filled in 84 minutes; and by the second and third in 140 minutes. What time will each pipe take to do it in? What time will be required, if the three pipes run together?

Call the contents of the cistern 1.

**115.]** EQUATIONS OF THE FIRST DEGREE.

Let $x$ denote the quantity discharged in 1 minute by the first;
$y$ " " " " " " second;
$z$ " " " " " " third;

then will $\dfrac{1}{x}$, $\dfrac{1}{y}$ and $\dfrac{1}{z}$ denote the number of minutes required for the pipes, separately, to fill the cistern; and,

$$\frac{1}{x+y+z}$$

will denote the number of minutes required for all three to fill it, running together;

from the conditions of the problem,

$$x + y = \frac{1}{70} \quad \cdots \quad (1)$$

$$x + z = \frac{1}{84} \quad \cdots \quad (2)$$

$$y + z = \frac{1}{140} \quad \cdots \quad (3);$$

clearing of fractions,

$$70x + 70y = 1 \quad \cdots \quad (4)$$
$$84x + 84z = 1 \quad \cdots \quad (5)$$
$$140y + 140z = 1 \quad \cdots \quad (6);$$

combining (1) and (2),

$$840y - 840z = 2 \quad \cdots \quad (7);$$

combining (6) and (7),

$$1680y = 8, \qquad \therefore \qquad y = \frac{1}{210};$$

substituting in (1), and transposing,

$$x = \frac{1}{70} - \frac{1}{210} = \frac{2}{210} = \frac{1}{105};$$

substituting in (3),

$$x = \frac{1}{140} - \frac{1}{210} = \frac{1}{420},$$

$$x + y + z = \frac{1}{210} + \frac{1}{105} + \frac{1}{420} = \frac{1}{60},$$

hence, $\frac{1}{x} = 105$, $\frac{1}{y} = 210$, $\frac{1}{z} = 420$, $\frac{1}{x+y+z} = 60$.

· 27. A has 3 purses, each containing a certain sum of money. If $20 be taken out of the first and put into the second, it will contain four times as much as remains in the first. If $60 be taken from the second and put into the third, then this will contain 1¾ times as much as there remains in the second. Again, if $40 be taken from the third and put into the first, then the third will contain 2⅞ times as much as the first. What were the contents of each purse?

Let $x$ denote the number of dollars in the first purse.

    $y$ " " " " second "

    $z$ " " " " third "

then from the conditions of the problem,

$$4(x - 20) = y + 20,$$
$$\tfrac{7}{4}(y - 60) = z + 60,$$
$$z - 40 = \tfrac{23}{8}(x + 40);$$

clearing of fractions, performing operations and transposing,

$$4x - y = 100 \ \ldots \ (1)$$
$$7y - 4z = 660 \ \ldots \ (2)$$
$$8z - 23x = 1240 \ \ldots \ (3);$$

combining (1) and (2),

$$28x - 4z = 1360 \ \ldots \ (4);$$

combining (3) and (4),

$$33x = 3960; \therefore x = 120;$$

by substitution, $y = 380; z = 500.$

28. A banker has two kinds of money; it takes $a$ pieces of the first to make a crown, and $b$ of the second to make the same sum. Some one offers him a crown for $c$ pieces. How many of each kind must the banker give him?

Since it takes $a$ pieces of the first to make 1 crown, $\dfrac{1}{a}$ = the part of a crown in each piece; and $\dfrac{1}{b}$, the part of a crown in each piece of the second:

let $x$ denote the number of pieces taken of the first kind,

$y$ " " " " second "

from the conditions of the problem,

$$x + y = c$$

$$\frac{x}{a} + \frac{y}{b} = 1, \text{ or } bx + ay = ab;$$

by combination,

$$by - ay = bc - ab; \text{ or } y(b - a) = b(c - a);$$

$$\therefore y = \frac{(a-c)b}{a-b}; \text{ whence, } x = \frac{a(c-b)}{a-b}.$$

29. Find what each of three persons, A, B, C, is worth, knowing, 1st, that what A is worth added to $l$ times what B and C are worth is equal to $p$; 2d, that what B is worth added to $m$ times what A and C are worth, is equal to $q$; 3d, that what C is worth added to $n$ times what A and B are worth, is equal to $r$.

Let $x$ denote what A is worth,
$y$ " " B "
$z$ " " C "

then, from the expressed conditions,

$$x + l\,(y + z) = p \quad \cdots \quad (1)$$
$$y + m\,(x + z) = q \quad \cdots \quad (2)$$
$$z + n\,(x + y) = r \quad \cdots \quad (3);$$

which, by adding and subtracting $lx$, $my$ and $nz$, may be written under the forms

$$(1 - l)\,x + l\,(x + y + z) = p \quad \cdots \quad (4)$$
$$(1 - m)\,y + m\,(x + y + z) = q \quad \cdots \quad (5)$$
$$(1 - n)\,z + n\,(x + y + z) = r \quad \cdots \quad (6);$$

dividing both members of each equation by the co-efficient of its first term,

$$x + \frac{l}{1 - l}\,(x + y + z) = \frac{p}{1 - l} \quad \cdots \quad (7)$$
$$y + \frac{m}{1 - m}\,(x + y + z) = \frac{q}{1 - m} \quad \cdots \quad (8)$$
$$z + \frac{n}{1 - n}\,(x + y + z) = \frac{r}{1 - n} \quad \cdots \quad (9);$$

adding these, member to member, and deducing from the resulting equation the value of $x + y + z$,

$$x + y + z = \frac{\dfrac{p}{1 - l} + \dfrac{q}{1 - m} + \dfrac{r}{1 - n}}{1 + \dfrac{l}{1 - l} + \dfrac{m}{1 - m} + \dfrac{n}{1 - n}} \quad \cdots \quad (10).$$

Denote the second member of equation (10) by the single letter $s$, a known quantity. Then by substituting this for the factor $x + y + z$, in each of the equations (7), (8) and (9), and deducing the values of $x$, $y$ and $z$, we have,

$$x = \frac{p}{1-l} - \frac{ls}{1-l} = \frac{p-ls}{1-l} \quad \cdots \quad (11)$$

$$y = \frac{q}{1-m} - \frac{ms}{1-m} = \frac{q-ms}{1-m} \quad \cdots \quad (12)$$

$$z = \frac{r}{1-n} - \frac{ns}{1-n} = \frac{r-ns}{1-n} \quad \cdots \quad (13).$$

Had we represented the polynomial $x + y + z$ by $s$, the algebraic work would be slightly diminished, but the preceding method has been followed in order to show more clearly the process of solution.

30. Find the values of the estates of six persons, A, B, C, D, E, F, from the following conditions: 1st. The sum of the estates of A and B is equal to $a$; that of C and D is equal to $b$; and that of E and F is equal to $c$. 2d. The estate of A is worth $m$ times that of C; the estate of D is worth $n$ times that of E, and the estate of F is worth $p$ times that of B.

*1st Solution.*

Let $x$ denote the value of A's estate;
$a - x$ " " " B's "
$y$ " " " C's "
$b - y$ " " " D's "
$z$ " " " E's "
$c - z$ " " " F's "

from the conditions of the problem,

$$\left.\begin{array}{l}x = my \\ b - y = nz \\ c - z = p(a-x)\end{array}\right\} \text{ or, } \begin{cases} x - my = 0 \cdot\cdot\cdot\cdot\cdot\cdot (1) \\ y + nz = b \cdot\cdot\cdot\cdot\cdot (2) \\ px - z = ap - c \cdot\cdot (3) \end{cases}$$

combining (1) and (2),
$$x + mnz = bm \cdot\cdot\cdot\cdot (4);$$
combining (3) and (4), and finding the value of $z$,
$$z = \frac{bmp + c - ap}{mnp + 1}, \text{ which denote by } P;$$
combining (3) and (4), and finding the value of $x$,
$$x = \frac{amnp + bm - cmn}{mnp + 1}, \text{ which denote by } Q;$$
substituting the last value in (1), and finding the value of $y$,
$$y = \frac{anp + b - cn}{mnp + 1} \text{ which denote by } R;$$
whence,

A's estate is equal to $\quad Q$,

B's " " " $a - Q$,

C's " " " $R$,

D's " " " $b - R$,

E's " " " $P$,

F's " " " $c - P$.

We might have combined (2) and (3), eliminating $z$, and then from this resulting equation, taken with (1), have found the value of $x$ and $y$.

### 2d. *Solution*

By a single unknown quantity,

Let $\quad x \quad$ denote the value of C's estate;

then will $b - x \quad$ " " D's "

$mx$      denote the value of A's  "
$a - mx$   "         "       B's .  "
$p(a - mx)$  "       "       F's    "
$c - p(a - mx)$ "    "       E's    "

and from the remaining condition of the problem,

$$b - x = n[c - p(a - mx)];$$

whence, by the rule for solving equations of the first degree,

$$x = \frac{b + anp - nc}{mnp + 1}.$$

Having found the value of C's estate, the remaining quantities may be found by substituting it in the expressions of the data, and reducing. The operations are obvious.

## INEQUALITIES.

1. Given,   $5x - 6 > 19$,   to find the smallest limit of $x$.

If we add 6 to both numbers of the inequality, we have

$$5x > 19 + 6, \quad \text{or,} \quad 5x > 25;$$

dividing both numbers by 5, we have

$$x > 5.$$

2. Given,   $3x + \frac{14}{2}x - 30 > 10$,   to find the least limit of $x$.

Reducing, $3x + 7x - 30 > 10$, or, $10x - 30 > 10$;

adding 30 to both members of the inequality, and dividing by 10, we have,

$$x > 4.$$

3. Given,   $\frac{x}{6} - \frac{x}{3} + \frac{x}{2} + \frac{13}{2} > \frac{17}{2}$,  to find the least limit of $x$.

Multiplying both members by 6, we have,

$$x - 2x + 3x + 39 > 51;$$

reducing, subtracting 39 from both members, and dividing by 2, we have,

$$x > 6.$$

4. Given, $\dfrac{ax}{5} + bx - ab > \dfrac{a^2}{5}$ to find the least limit of $x$.

Multiplying both members by 5, we have

$$ax + 5bx - 5ab > a^2;$$

adding $+ 5ab$ to both members, and dividing by the co-efficients of $x$, we have,

$$(a + 5b)x > a(a + 5b); \quad \text{or,} \quad x > a.$$

5. Given, $\dfrac{bx}{7} - ax + ab < \dfrac{b^2}{7}$, to find the largest limit of $x$.

Multiplying both members by 7, adding $-7ab$, and dividing by the co-efficients of $x$, we have,

$$(b - 7a)x < b(b - 7a) \quad \text{or,} \quad x < b.$$

## REDUCTION OF RADICALS.

1. $\dfrac{7}{3 - \sqrt{5}}$. Multiply both terms by $3 + \sqrt{5}$.

2. $\dfrac{7\sqrt{5}}{\sqrt{11} + \sqrt{3}}$. Multiply both terms by $\sqrt{11} - \sqrt{3}$.

3. $\dfrac{3 + 2\sqrt{7}}{5\sqrt{12} - 6\sqrt{5}}$. Multiply both terms by $5\sqrt{12} + 6\sqrt{5}$.

4. $\dfrac{(3+\sqrt{3})(3+\sqrt{5})(\sqrt{5}-2)}{(5-\sqrt{5})(1+\sqrt{3})}$. Multiply both terms by $5+\sqrt{5}$ and $1-\sqrt{3}$; this gives,

$$\dfrac{(3+\sqrt{3})(1-\sqrt{3})(3+\sqrt{5})(5+\sqrt{5})(\sqrt{5}-2)}{(5-\sqrt{5})(5+\sqrt{5})(1+\sqrt{3})(1-\sqrt{3})};$$

which, after performing the operations indicated, and reducing, becomes,

$$\dfrac{-40\sqrt{15}-80\sqrt{3}+80\sqrt{3}+32\sqrt{15}}{20x-2}=\dfrac{1}{5}\sqrt{15}.$$

5. $\dfrac{\sqrt{a+x}+\sqrt{a-x}}{\sqrt{a+x}-\sqrt{a-x}}$. Multiplying both terms by $\sqrt{a+x}+\sqrt{a-x}$, we have,

$$\dfrac{2a+2\sqrt{a^2-x^2}}{2x}=\dfrac{a+\sqrt{a^2-x^2}}{x}.$$

6. $\dfrac{(7-2\sqrt{5})+2+\sqrt{45}}{2\sqrt{3}-\sqrt{7}}$. Multiplying both terms by $2\sqrt{3}+\sqrt{7}$, the denominator becomes 5. For the numerator, performing the multiplications and reductions, we have,

$$\begin{array}{l}7-2\sqrt{5}+2+\sqrt{45}\\ 2\sqrt{3}+\sqrt{7}\\ \hline 14\sqrt{3}-4\sqrt{15}+4\sqrt{3}+2\sqrt{135}\\ \quad +7\sqrt{7}-2\sqrt{35}+2\sqrt{7}+\sqrt{315}.\end{array}$$

But, $2\sqrt{135}=6\sqrt{15}$; and $\sqrt{315}=3\sqrt{35}$; hence, the product reduces to

$$18\sqrt{3}+2\sqrt{15}+\sqrt{35}+9\sqrt{7}.$$

But,
$$18\sqrt{3} = 9 \times 2\sqrt{3}; \text{ and } 2\sqrt{15} = 2\sqrt{3} \times \sqrt{5};$$
also,
$$\sqrt{35} = \sqrt{7} \times \sqrt{5};$$
hence, the sum reduces to
$$9 \times 2\sqrt{3} + 2\sqrt{3} \times \sqrt{5} + \sqrt{7} \times \sqrt{5} + 9\sqrt{7}$$
$$= 9(2\sqrt{3} + \sqrt{7}) + \sqrt{5}(2\sqrt{3} + \sqrt{7})$$
$$= (9 + \sqrt{5})(2\sqrt{3} + \sqrt{7}).$$

## EQUATIONS OF THE SECOND DEGREE.

7. Given $\quad \dfrac{x^2}{3} - \dfrac{a}{b}x = 1 - \dfrac{b}{a}x - \dfrac{2x^3}{3},\quad$ to find the values of $x$.

Clearing of fractions, transposing and reducing,
$$x^2 - \frac{a^2 - b^2}{ab}x = 1;$$
whence, by the rule
$$x = \frac{a^2 - b^2}{2ab} \pm \sqrt{1 + \frac{a^4 - 2a^2b^2 + b^4}{4a^2b^2}} = \frac{a^2 - b^2}{2ab} \pm \frac{a^2 + b^2}{2ab};$$

taking the upper sign, $\quad x = \dfrac{2a^2}{2ab} = \dfrac{a}{b},$

" lower " $\quad x = -\dfrac{2b^2}{2ab} = -\dfrac{b}{a}.$

8 Given $\quad \dfrac{dx}{c} + \dfrac{3x^2}{4} + 1 = \dfrac{1+c}{c} - \dfrac{x^2}{4} + \dfrac{x}{d},$

to find the values of $x$.

Clearing of fractions, transposing and reducing,
$$x^2 + \frac{d^2 - c}{cd}x = \frac{1}{c},$$

**159.]** EQUATIONS OF THE SECOND DEGREE. 75

whence, by the rule,

$$x = -\frac{d^2-c}{2cd} \pm \sqrt{\frac{1}{c} + \frac{d^4 - 2cd^2 + c^2}{4c^2d^2}} = -\frac{d^2-c}{2cd} \pm \frac{d^2+c}{2cd};$$

taking the upper sign, $\quad x = \frac{2c}{2cd} = \frac{1}{d},$

" lower " $\quad x = -\frac{2d^2}{2cd} = -\frac{d}{c}.$

9. Given $\dfrac{x^2}{4} - \dfrac{2x}{3} + \dfrac{59}{8} = 8 - \dfrac{x^2}{4} - \dfrac{x}{3},$ to find the values of $x$.

Clearing of fractions, transposing and reducing,

$$x^2 - \frac{2}{3}x = \frac{5}{4},$$

whence, by the rule,

$$x = \frac{1}{3} \pm \sqrt{\frac{5}{4} + \frac{1}{9}} = \frac{1}{3} \pm \frac{7}{6};$$

hence, $\quad x = \dfrac{3}{2} \quad$ and $\quad x = -\dfrac{5}{6}.$

10. Given $\dfrac{90}{x} - \dfrac{90}{x+1} = \dfrac{27}{x+2},$ to find the values of $x$.

Dividing both members by 9,

$$\frac{10}{x} - \frac{10}{x+1} = \frac{3}{x+2};$$

clearing of fractions, transposing and reducing,

$$x^2 - \frac{7}{3}x = \frac{20}{3};$$

whence, by the rule,

$$x = \frac{7}{6} \pm \sqrt{\frac{20}{3} + \frac{49}{36}} = \frac{7}{6} \pm \frac{17}{6};$$

hence, $\quad x = 4,\quad$ and $\quad x = -\dfrac{10}{6} = -\dfrac{5}{3}.$

11. Given $\quad \dfrac{2x-10}{8-x} - 2 = \dfrac{x+3}{x-2},\;$ to find the values of $x$.

Clearing of fractions, &c.,

$$x^2 - \frac{39}{5}x = -\frac{28}{5};$$

whence by the rule,

$$x = \frac{39}{10} \pm \sqrt{-\frac{28}{5} + \frac{1521}{100}} = \frac{39}{10} \pm \frac{31}{10};$$

hence, $\quad x = 7,\quad$ and $\cdot\; x = \dfrac{8}{10} = \dfrac{4}{5}.$

12. Given $\quad ax - \dfrac{x^2}{b} + b = \dfrac{b-1}{b}x^2 + \dfrac{b}{a}x,\;$ to find the values of $x$.

Clearing of fractions, &c.,

$$x^2 - \frac{a^2-b}{a}x = b;$$

whence, by the rule,

$$x = \frac{a^2-b}{2a} \pm \sqrt{b + \frac{a^4 - 2a^2b + b^2}{4a^2}} = \frac{a^2-b}{2a} \pm \frac{a^2+b}{2a};$$

hence, $\quad x = a,\quad x = -\dfrac{b}{a}.$

13. Given $\quad \dfrac{a-b}{c}x + \dfrac{3x^2}{2} - \dfrac{a^2}{c^2} = \dfrac{b+a}{c}x + \dfrac{x^2}{2} - \dfrac{b^2}{c^2},$
to find the values of $x$.

Clearing of fractions, &c.,

$$x^2 - \frac{2b}{c}x = \frac{a^2 - b^2}{c^2};$$

whence, by the rule,

$$x = \frac{b}{c} \pm \sqrt{\frac{a^2 - b^2}{c^2} + \frac{b^2}{c^2}} = \frac{b}{c} \pm \frac{a}{c}$$

hence, $\quad x = \dfrac{b+a}{c}, \quad \text{and} \quad x = \dfrac{b-a}{c}.$

14. Given $\quad mx^2 + mn = 2m\sqrt{n} \cdot x + nx^2$, to find the values of $x$.

Transposing and reducing,

$$x^2 - \frac{2m\sqrt{n}}{m-n}x = -\frac{mn}{m-n};$$

whence, by the rule,

$$x = \frac{m\sqrt{n}}{m-n} \pm \sqrt{-\frac{mn}{m-n} + \frac{m^2 n}{(m-n)^2}} = \frac{m\sqrt{n}}{m-n} \pm \frac{n\sqrt{m}}{m-n};$$

hence, $x = \dfrac{m\sqrt{n} + n\sqrt{m}}{m-n} = \dfrac{\sqrt{mn}(\sqrt{m}+\sqrt{n})}{(\sqrt{m}+\sqrt{n})(\sqrt{m}-\sqrt{n})} = \dfrac{\sqrt{mn}}{\sqrt{m}-\sqrt{n}}$

since, $\quad m - n = (\sqrt{m} + \sqrt{n})(\sqrt{m} - \sqrt{n}).\quad$ (Art. 47.)

also,

$$x = \frac{m\sqrt{n} - n\sqrt{m}}{m-n} = \frac{\sqrt{mn}(\sqrt{m}-\sqrt{n})}{(\sqrt{m}+\sqrt{n})(\sqrt{m}-\sqrt{n})} = \frac{\sqrt{mn}}{\sqrt{m}+\sqrt{n}}.$$

*2d Solution.*

$$mx^2 + mn = 2m\sqrt{n} \cdot x + nx^2;$$

transposing the first term of the second member, we have,

$$mx^2 - 2m\sqrt{n} \cdot x + mn = nx^2;$$

observing that the first member is the square of a binomial whose terms are $\sqrt{m} \cdot x - \sqrt{m}\sqrt{n}$, we have,

$$\sqrt{m} \cdot x - \sqrt{m}\sqrt{n} = \pm \sqrt{n} \cdot x;$$

$$\sqrt{m} \cdot x \mp \sqrt{n} \cdot x = \sqrt{m}\sqrt{n}$$

$$(\sqrt{m} \mp \sqrt{n})x = \sqrt{m}\sqrt{n}: \qquad \text{hence,}$$

$$x = \frac{\sqrt{mn}}{\sqrt{m} - \sqrt{n}} \quad \text{and,} \quad x = \frac{\sqrt{mn}}{\sqrt{m} + \sqrt{n}}.$$

15. Given $\quad abx^2 - \dfrac{6a^2}{c^2} + \dfrac{b^2 x}{c} = \dfrac{ab - 2b^2}{c^2} - \dfrac{3a^2}{c}x,$

to find the values of $x$.

Clearing of fractions, &c.,

$$x^2 + \frac{b^2 + 3a^2}{abc}x = \frac{ab - 2b^2 + 6a^2}{abc^2};$$

whence, by the rule,

$$x = -\frac{b^2 + 3a^2}{2abc} \pm \sqrt{\frac{ab - 2b^2 + 6a^2}{abc^2} + \frac{b^4 + 6a^2 b^2 + 9a^4}{4a^2 b^2 c^2}}$$

$$= -\frac{b^2 + 3a^2}{2abc} \pm \frac{3a^2 + 4ab - b^2}{2abc};$$

hence, $x = \dfrac{4ab - 2b^2}{2abc} = \dfrac{2a - b}{ac}, \quad x = -\dfrac{6a^2 + 4ab}{2abc} = -\dfrac{3a + 2b}{bc}$

16. Given $\quad \dfrac{4x^2}{7} + \dfrac{2x}{7} + 10 = 19 - \dfrac{3x^2}{7} + \dfrac{58x}{7},$

to find the values of $x$.

## EQUATIONS OF THE SECOND DEGREE.

Clearing of fractions, &c.,

$$x^2 - 8x = 9;$$

whence, by the rule,

$$x = 4 \pm \sqrt{9 + 16} = 4 \pm 5;$$

hence, $\quad x = 9, \quad x = -1.$

17. Given $\dfrac{x+a}{x-a} - b = \dfrac{a-x}{a+x}$, to find the values of $x$.

Clearing of fractions, &c.,

$$x^2 = \frac{(b+2)a^2}{b-2};$$

whence, by extracting the square root of both members,

$$x = + a \sqrt{\frac{b+2}{b-2}}, \quad \text{and} \quad x = -a \sqrt{\frac{b+2}{b-2}}$$

18. Given $\quad 2x + 2 = 24 - 5x - 2x^2$, to find the values of $x$.

Transposing and reducing,

$$x^2 + \frac{7}{2}x = 11;$$

whence, by the rule,

$$x = -\frac{7}{4} \pm \sqrt{11 + \frac{49}{16}} = -\frac{7}{4} \pm \frac{15}{4};$$

whence, $\quad x = 2, \quad \text{and} \quad x = -\dfrac{11}{2}.$

19. Given $\quad x^2 - x - 40 = 170$, to find the values of $x$.

Transposing,

$$x^2 - x = 210;$$

whence, by the rule,

$$x = \frac{1}{2} \pm \sqrt{210 + \frac{1}{4}} = \frac{1}{2} \pm \frac{29}{2};$$

hence, $\quad x = 15, \quad$ and $\quad x = -14.$

20. Given $\quad 3x^2 + 2x - 9 = 76,\quad$ to find the values of $x$.

Transposing and reducing,

$$x^2 + \frac{2}{3}x = \frac{85}{3};$$

whence, by the rule,

$$x = -\frac{1}{3} \pm \sqrt{\frac{85}{3} + \frac{1}{9}} = -\frac{1}{3} \pm \frac{16}{3};$$

hence, $\quad x = 5, \quad$ and $\quad x = -\dfrac{17}{3} = -5\dfrac{2}{3}.$

21. Given $\quad a^2 + b^2 - 2bx + x^2 = \dfrac{m^2 x^2}{n^2},\quad$ to find the values of $x$

Clearing of fractions, &c.,

$$x^2 + \frac{2bn^2}{m^2 - n^2}x = \frac{n^2 a^2 + n^2 b^2}{m^2 - n^2};$$

whence, by the rule,

$$x = -\frac{bn^2}{m^2 - n^2} \pm \sqrt{\frac{n^2 a^2 + n^2 b^2}{m^2 - n^2} + \frac{b^2 n^4}{m^4 - 2m^2 n^2 + n^4}}$$

$$= \frac{bn^2}{n^2 - m^2} \pm \frac{n\sqrt{a^2 m^2 + b^2 m^2 - a^2 n^2}}{n^2 - m^2};$$

whence, $\quad x = \dfrac{n}{n^2 - m^2}\left\{bn \pm \sqrt{a^2 m^2 + b^2 m^2 - a^2 n^2}\right\}$

## EQUATIONS OF THE SECOND DEGREE.

22. Given $\quad \dfrac{1}{2(x-1)} + \dfrac{3}{x^2-1} = \dfrac{1}{4}$;

whence, $\quad 2(x+1) + 12 = x^2 - 1$,

and $\quad \therefore\ x = 1 \pm \sqrt{16},\ x' = 5,\ x'' = -3$.

23. Given $\quad \dfrac{x}{15} + \dfrac{40}{3(10-x)} = \dfrac{3(10+x)}{95}$;

whence, $\quad 19x(10-x) + 3800 = 9(10+x)(10-x)$.

Reducing, $\quad x^2 - 19x = 290$;

hence, $\quad x = \dfrac{19}{2} \pm \sqrt{290 + \dfrac{361}{4}} = \dfrac{19 \pm 39}{2}$.

$\therefore\ x' = 29,\ x'' = -10$.

24. Given $\quad \dfrac{x^2 - 5x}{x+3} = x - 3 + \dfrac{1}{x}$;

whence, $\quad x(x^2 - 5x) = (x^2 - 9)x + x + 3$.

Reducing, $\quad x^2 - \dfrac{8}{5}x = -\dfrac{3}{5}$;

hence, $\quad x = \dfrac{4}{5} \pm \sqrt{\dfrac{16}{25} - \dfrac{15}{25}}$;

or $\quad x = \dfrac{4 \pm 1}{5};\ x' = 1,\ x'' = \dfrac{3}{5}$;

25. Given $\quad \dfrac{x}{x-1} = \dfrac{3}{2} + \dfrac{x-1}{x}$;

whence, $\quad 2x^2 = 3(x^2 - x) + 2(x^2 - 2x + 1)$;

hence, $\quad x^2 - \dfrac{7}{3}x = -\dfrac{2}{3}$, or $\quad x = \dfrac{7}{6} \pm \sqrt{\dfrac{49}{36} - \dfrac{2}{3}}$;

$\therefore\ x' = 2 \quad \text{and} \quad x'' = \dfrac{1}{3}$.

26. Given $\quad \dfrac{x+2}{x-2} - \dfrac{x-2}{x+2} = \dfrac{5}{6}$;

whence, $\quad 6(x^2 + 4x + 4) - 6(x^2 - 4x + 4) = 5(x^2 - 4)$;

therefore, $\quad x^2 - \dfrac{48}{5}x = 4$;

and by the rule, $x = \dfrac{24}{5} \pm \sqrt{\dfrac{576}{25} + 4} = \dfrac{24 \pm 26}{5}$;

$$\therefore x' = 10 \quad \text{and} \quad x'' = -\dfrac{2}{5}.$$

27. Given $\dfrac{x-6}{x-12} - \dfrac{x-12}{x-6} = \dfrac{5}{6}$;

whence, $6(x^2 - 12x + 36) - 6(x^2 - 24x + 144) = 5(x^2 - 18x + 72)$

Reducing, $x^2 - \dfrac{162}{5}x = -\dfrac{1008}{5}$;

and $x = \dfrac{81}{5} \pm \sqrt{\dfrac{6561}{25} - \dfrac{5040}{25}} = \dfrac{81 \pm 39}{5}$;

$$\therefore x' = 24 \quad \text{and} \quad x'' = \dfrac{42}{5}.$$

28. Given $\dfrac{x}{x+1} + \dfrac{x+1}{x} = \dfrac{13}{6}$;

whence, $6x^2 + 6(x^2 + 2x + 1) = 13(x^2 + x)$;

hence, $x = -\dfrac{1}{2} \pm \sqrt{6\tfrac{1}{4}} = \dfrac{-1 \pm 5}{2}$;

$$\therefore x' = 2 \quad \text{and} \quad x'' = -3.$$

29. Given $\dfrac{1}{x-2} - \dfrac{2}{x+2} = \dfrac{3}{5}$;

whence, $5(x+2) - 10(x-2) = 3(x^2 - 4)$;

hence, $x = -\dfrac{5}{6} \pm \sqrt{14 + \dfrac{25}{36}} = -\dfrac{5 \pm 23}{6}$;

$$\therefore x' = 3 \quad \text{and} \quad x'' = -\dfrac{14}{3}.$$

30. Given $\dfrac{4}{x+1} + \dfrac{5}{x+2} = \dfrac{12}{x+3}$;

whence, $4(x^2 + 5x + 6) + 5(x^2 + 4x + 3) = 12(x^2 + 3x + 2)$;

hence, $x = \dfrac{2}{3} \pm \sqrt{\dfrac{4}{9} + \dfrac{45}{9}} = \dfrac{2 \pm 7}{3}$;

$$\therefore x' = 3 \quad \text{and} \quad x'' = -\dfrac{5}{3}.$$

PROBLEMS GIVING RISE TO EQUATIONS OF THE SECOND DEGREE.

4. A grazier bought as many sheep as cost him £60, and after reserving 15 out of the number, he sold the remainder for £54, and gained 2s. a head on those he sold: how many did he buy?

Let $x$ denote the number purchased:

and $x - 15$, the number sold;

then will $\dfrac{1200}{x}$ denote the number of shillings paid for 1 sheep,

and $\dfrac{1080}{x - 15}$ the number of shillings received for each.

From the conditions of the problem,

$$\frac{1200}{x} = \frac{1080}{x - 15} - 2;$$

clearing of fractions, &c.,

$$x^2 + 45x = 9000:$$

whence, by the rule,

$$x = -\frac{45}{2} \pm \sqrt{9000 + \frac{2025}{4}} = -\frac{45}{2} \pm \frac{195}{2};$$

hence, $x = 75$, and $x = -120$,

the positive value only, corresponds to the required solution.

5. A merchant bought cloth for which he paid £33 15s., which he sold again at £2 8s. per piece, and gained by the bargain as much as one piece cost him: how many pieces did he buy?

Let $x$ denote the number of pieces purchased:

then will, $\dfrac{675}{x}$ denote he number of shillings paid for each, and $48x$ the number of shillings for which he sold the whole.

From the conditions of the problem,

$$48x - 675 = \frac{675}{x};$$

then, by clearing of fractions, &c.,

$$x^2 - \frac{225}{16}x = \frac{225}{16}.$$

whence, by the rule,

$$x = \frac{225}{32} \pm \sqrt{\frac{225}{16} + \frac{50625}{1024}} = \frac{225}{32} \pm \frac{255}{32};$$

using the positive value only, $\quad x = \dfrac{480}{32} = 15.$

6. What number is that, which being divided by the product of its digits, the quotient will be 3; and if 18 be added to it, the order of its digits will be reversed?

Let $x$ denote the first digit,

and, $y$ " second "

then will $\quad\quad\quad\quad\quad 10x + y \quad\quad$ denote the number.

From the conditions of the problem,

$$\frac{10x + y}{xy} = 3,$$

$$10x + y + 18 = 10y + x;$$

whence, by reduction,

$$10x + y = 3xy,$$

$$y - x = 2;$$

finding the value of $x$ in terms of $y$ from the second, and substituting in the first, we have,

$$10y - 20 + y = 3y^2 - 6y;$$

whence, by transposing, &c.,

$$y^2 - \frac{17}{3}y^2 = -\frac{20}{3}; \quad \text{and,}$$

by the rule,

$$y = +\frac{17}{6} \pm \sqrt{-\frac{20}{3} + \frac{289}{36}} = +\frac{17}{6} \pm \frac{7}{6},$$

taking the positive sign, $\quad y = 4;$

whence, $\quad z = 2,\ $ and the number is 24.

7. Find a number such that if you subtract it from 10, and multiply the remainder by the number itself, the product will be 21.

Let $x$ denote the number:

from the conditions of the problem,

$$(10 - x)\,x = 21; \quad \text{or,} \quad x^2 - 10x = -21;$$

by the rule,

$$x = 5 \pm \sqrt{-21 + 25} = 5 \pm 2;$$

whence, $\quad x = 7,\ $ and $\ x = 3.$

8. Two persons, A and B, departed from different places at the same time, and travelled towards each other. On meeting, it appeared that A had travelled 18 miles more than B; and that A could have performed B's journey in $15\frac{3}{4}$ days, but B would have been 28 days in performing A's journey. How far did each travel?

Let $\quad x \quad$ denote the number of miles B travelled;

$x + 18$ " " " A "

$\dfrac{x}{15\frac{3}{4}}$ " " " A " in one day;

$\dfrac{x+18}{28}$ denote the number of miles B travelled in one day,

$\dfrac{x+18}{\left(\dfrac{x}{15\frac{3}{4}}\right)}$  "        "        days A  "

$\dfrac{x}{\left(\dfrac{x+18}{28}\right)}$  "        "        " B  "

from the conditions of the problem,

$$\dfrac{x+18}{\dfrac{x}{15\frac{3}{4}}} = \dfrac{x}{\dfrac{x+18}{28}} \quad \text{or,} \quad \dfrac{x^2 + 36x + 324}{28} = \dfrac{x^2}{15\frac{3}{4}};$$

clearing of fractions, and reducing,

$$x^2 - \dfrac{324}{7}x = \dfrac{2916}{7}.$$

By the rule,

$$x = \dfrac{162}{7} \pm \sqrt{\dfrac{2916}{7} + \dfrac{26244}{49}}.$$

$$x = \dfrac{162}{7} \pm \dfrac{216}{7};$$

hence, using the upper sign,

$$x = \dfrac{378}{7} = 54.$$

9. A company at a tavern had £8 15s. to pay for their reckoning; but before the bill was settled, two of them left the room, and then those who remained had 10s. apiece more to pay than before: how many were there in the company?

Let $x$ denote the number in the company.

Then $\dfrac{175}{x}$ will denote the number of shillings each should pay;

$\dfrac{175}{x-2}$ " " " " paid;

from the conditions of the problem,

$$\frac{175}{x-2} - \frac{175}{x} = 10;$$

clearing of fractions,

$$175x - 175x + 350 = 10x^2 - 20x;$$

whence, $\qquad x^2 - 2x = 35.$

By the rule,

$$x = 1 \pm \sqrt{36} = 1 \pm 6;$$

using the upper sign, $\quad x = 7.$

10. What two numbers are those whose difference is 15, and of which the cube of the lesser is equal to half their product?

Let $x$ denote the smaller number;
then will $x + 15$ " greater "
from the conditions of the problem,

$$x^3 = \frac{1}{2}(x^2 + 15x), \quad \text{or}, \quad x^2 = \frac{1}{2}(x + 15);$$

whence, $\qquad x^2 - \dfrac{1}{2}x = \dfrac{15}{2}.$

By the rule,

$$x = \frac{1}{4} \pm \sqrt{\frac{15}{2} + \frac{1}{16}} = \frac{1}{4} \pm \frac{11}{4};$$

using the upper sign, $\quad x = 3;$ hence, $\quad x + 15 = 18.$

11. Two partners, A and B, gained $140 in trade: A's money

was 3 months in trade, and his gain was $60 less than his stock: B's money was $50 more than A's, and was in trade 5 months: what was A's stock?

Let $x$ denote the number of dollars in A's stock;

$x + 50$ " " " B's "

$x - 60$ " A's total gain;

$\dfrac{x-60}{3}$ " A's gain per month;

$\dfrac{x-60}{3x}$ " A's " " " on 1 dollar;

$\left(\dfrac{x-60}{3x}\right)(x+50)$ B's " " "

$\left(\dfrac{x-60}{3x}\right)(x+50)\,5$ B's total gain.

From the conditions of the problem,

$$x - 60 + \dfrac{(x-60)}{3x}(x+50)\,5 = 140\,;$$

clearing of fractions, and reducing,

$$x^2 - \dfrac{325}{4}x = 1875.$$

By the rule,

$$x = \dfrac{325}{8} \pm \sqrt{+1875 + \dfrac{105625}{64}} = \dfrac{325}{8} \pm \dfrac{475}{8}\,;$$

whence, $x = 100$.

12. Two persons, A and B, start from two different points, and travel toward each other. When they meet, it appears that A has travelled 30 miles more than B. It also appears that it will take A 4 days to travel the road that B had come, and B 9 days to travel

the road which A had come. What was their distance apart when they set out?

Let $x$ denote the number of miles B travelled;
then will $x + 30$ " " " A "

$\dfrac{x}{4}$ " " " A travels per day;

$\dfrac{x+30}{9}$ " " " B " "

$\dfrac{x+30}{\left(\dfrac{x}{4}\right)}$ " " days A "

$\dfrac{x}{\left(\dfrac{x+30}{9}\right)}$ " " " B "

From the conditions,

$$\dfrac{x}{\left(\dfrac{x+30}{9}\right)} = \dfrac{x+30}{\left(\dfrac{x}{4}\right)} \quad \text{or,} \quad \dfrac{x^2}{4} = \dfrac{x^2 + 60x + 900}{9};$$

whence, by reduction,

$$x^2 - 48x = 720 \cdot$$

and by the rule,

$$x = 24 \pm \sqrt{720 + 576} = 24 \pm 36;$$

taking the upper sign, $x = 60$, and $x + 30 = 90$;
hence, the distance is 150 miles.

### EXAMPLES INVOLVING RADICALS OF THE SECOND DEGREE.

3. Given $\dfrac{a}{x} + \sqrt{\dfrac{a^2 - x^2}{x^2}} = \dfrac{x}{b}$, to find the values of $x$.

Multiplying both members by $bx$, and transposing,

$$b\sqrt{a^2 - x^2} = x^2 - ab;$$

squaring both members,

$$b^2a^2 - b^2x^2 = x^4 \cdot\cdot 2abx^2 + a^2b^2;$$

cancelling $b^2a^2$, dividing both members by $x^2$ and transposing,

$$x^2 = 2ab - b^2 \quad \therefore \quad x = \pm \sqrt{2ab - b^2}.$$

4. Given $\sqrt{\dfrac{x+a}{x}} + 2\sqrt{\dfrac{a}{x+a}} = b^2 \sqrt{\dfrac{x}{x+a}}$, to find $x$

Multiplying both members by $\sqrt{\dfrac{x+a}{x}}$, .

$$\frac{x+a}{x} + 2\sqrt{\frac{a}{x}} = b^2;$$

multiplying both members by $x$, and transposing,

$$2\sqrt{ax} = b^2x - x - a = (b^2 - 1)x - a;$$

squaring both members,

$$4ax = (b^4 - 2b^2 + 1)x^2 - 2a(b^2 - 1)x + a^2;$$

transposing and reducing,

$$x^2 - \frac{2a(b^2 + 1)}{b^4 - 2b^2 + 1}x = -\frac{a^2}{b^4 - 2b^2 + 1},$$

whence,

$$x = \frac{a(b^2 + 1)}{b^4 - 2b^2 + 1} \pm \sqrt{-\frac{a^2}{b^4 - 2b^2 + 1} + \frac{a^2(b^4 + 2b^2 + 1)}{(b^4 - 2b^2 + 1)^2}},$$

or,

$$x = \frac{a(b^2 + 1)}{b^4 - 2b^2 + 1} \pm \frac{2ab}{b^4 - 2b^2 + 1};$$

now,

$$b^4 - 2b^2 + 1 = (b - 1)^2(b + 1)^2.$$

## EQUATIONS OF THE SECOND DEGREE.

Hence, taking the upper sign and reducing,

$$x = \frac{a(b+1)^2}{(b^2-1)^2} = \frac{a(b+1)^2}{(b-1)^2(b+1)^2} = \frac{a}{(b-1)^2};$$

and taking the lower sign and reducing,

$$x = \frac{a(b-1)^2}{(b^2-1)^2} = \frac{a(b-1)^2}{(b-1)^2(b+1)^2} = \frac{a}{(b+1)^2},$$

or, uniting the two values in a single expression,

$$x = \frac{a}{(b \mp 1)^2}$$

5. Given, $\dfrac{a - \sqrt{a^2 - x^2}}{a + \sqrt{a^2 - x^2}} = b,$ to find $x$.

Clearing of fractions, transposing, &c.,

$$a(1-b) = (b+1)\sqrt{a^2 - x^2};$$

squaring both members,

$$a^2(1-b)^2 = (b+1)^2(a^2 - x^2);$$

transposing and reducing,

$$x^2 = \frac{4a^2 b}{(b+1)^2} \qquad \therefore \qquad x = \pm \frac{2a\sqrt{b}}{b+1}.$$

6. Given, $\dfrac{\sqrt{x} + \sqrt{x-a}}{\sqrt{x} - \sqrt{x-a}} = \dfrac{n^2 a}{x-a},$ to find $x$.

Multiplying both terms of the first member by $\sqrt{x} - \sqrt{x-a}$, and then dividing both members by $a$,

$$\frac{1}{2x - a - 2\sqrt{x^2 - ax}} = \frac{n^2}{x - a};$$

clearing of fractions, &c.,

$$(1 - 2n^2)x - (1 - n^2)a = -2n^2 \sqrt{x^2 - ax};$$

squaring both members, transposing and reducing,

$$x^2 - \frac{2(1 - 3n^2)a}{1 - 4n^2} x = -\frac{1 - 2n^2 + n^4}{1 - 4n^2} a^2;$$

whence by the rule,

$$x = \frac{1 - 3n^2}{1 - 4n^2} a \pm \sqrt{-\frac{(1 - 2n^2 + n^4)a^2}{1 - 4n^2} + \frac{(1 - 6n^2 + 9n^4)a^2}{(1 - 4n^2)^2}}$$

$$x = \frac{(1 - 3n^2)}{1 - 4n^2} a \pm \frac{2n^3 a}{1 - 4n^2} = \frac{(1 - 3n^2 \pm 2n^3)a}{1 - 4n^2}.$$

Taking the upper sign, and dividing both terms of the fraction by

$$1 + 2n,$$

$$x = \frac{(1 - 2n + n^2)a}{1 - 2n} = \frac{(1 - n)^2 a}{1 - 2n}.$$

Taking the lower sign, and dividing both terms by $1 - 2n$,

$$x = \frac{a(1 + 2n + n^2)}{1 + 2n} = \frac{(1 + n)^2 a}{1 + 2n},$$

taking the two values together, $x = \dfrac{a(1 \pm n)^2}{1 \pm 2n}$ . .

7. Given $\dfrac{\sqrt{a + x}}{\sqrt{x}} + \dfrac{\sqrt{a - x}}{\sqrt{x}} = \sqrt{\dfrac{x}{b}}$, to find $x$.

Multiplying both members by $\sqrt{x}$,

$$\sqrt{a + x} + \sqrt{a - x} = \frac{x}{\sqrt{b}};$$

squaring both members,

$$2a + 2\sqrt{a^2 - x^2} = \frac{x^2}{b}; \quad \text{or,} \quad 2\sqrt{a^2 - x^2} = \frac{x^2}{b} - 2a;$$

squaring both members,
$$4(a^2 - x^2) = \frac{x^4}{b^2} - \frac{4ax^2}{b} + 4a^2;$$
cancelling $4a^2$ in the two members, and dividing both by $x^2$,
$$-4 = \frac{x^2}{b^2} - \frac{4a}{b};$$
clearing of fractions, &c.,
$$x^2 = 4ab - 4b^2 \quad \therefore \quad x = \pm 2\sqrt{ab - b^2}.$$

8. Given $\quad \dfrac{a + x + \sqrt{2ax + x^2}}{a + x} = b$; to find $x$.

Clearing of fractions, transposing and factoring,
$$\sqrt{2ax + x^2} = (b - 1)(a + x).$$
Squaring both members,
$$2ax + x^2 = (b^2 - 2b + 1)(a^2 + 2ax + x^2); \quad \text{or,}$$
$$2ax + x^2 = (b^2 - 2b)(a^2 + 2ax + x^2) + a^2 + 2ax + x^2;$$
whence, by reduction,
$$x^2 + 2ax = -a^2 \left(\frac{1 - 2b + b^2}{b^2 - 2b}\right);$$
whence, by the rule,
$$x = -a \pm \sqrt{-a^2\left(\frac{1 - 2b + b^2}{b^2 - 2b}\right) + a^2} = -a \pm a \frac{1}{\sqrt{2b - b^2}};$$
taking the upper sign, $\quad x = \dfrac{a(1 - \sqrt{2b - b^2})}{\sqrt{2b - b^2}};$

taking the lower sign, $\quad x = -\dfrac{a(1 + \sqrt{2b - b^2})}{\sqrt{2b - b^2}};$

whence, $\quad x = \pm \dfrac{a(1 \mp \sqrt{2b - b^2})}{\sqrt{2b - b^2}}.$

## 2d Solution.

Make $x + a = y$; whence, $2ax + x^2 = y^2 - a^2$

substituting in the equation, and clearing of fractions,

$$y + \sqrt{y^2 - a^2} = by;$$

transposing, &c.,

$$\sqrt{y^2 - a^2} = (b - 1)y;$$

squaring both members, and cancelling,

$$- a^2 = (b^2 - 2b)y^2;$$

whence, solving with respect to $y$,

$$y = \pm \frac{a}{\sqrt{2b - b^2}};$$

substituting for $y$ its value, &c.,

$$x = -a \pm \frac{a}{\sqrt{2b - b^2}};$$

whence, as before, $\quad x = \dfrac{\pm a(1 \mp \sqrt{2b - b^2})}{\sqrt{2b - b^2}}$

## TRINOMIAL EQUATIONS.

6. Given $\quad x^4 - (2bc + 4a^2)x^2 = -b^2c^2$; to find $x$.

By the rule,

$$x = \pm \sqrt{bc + 2a^2 \pm \sqrt{4a^2bc + 4a^4}} = \pm\sqrt{bc + 2a^2 \pm 2a\sqrt{bc + a^2}}.$$

7. Given $\quad 2x - 7\sqrt{x} = 99$; or, $2x - 99 = 7\sqrt{x}$

Squaring both members

$$4x^2 - 396x + 9801 = 49x;$$

transposing and reducing,

$$x^2 - \frac{445}{4}x = -\frac{9801}{4}$$

whence, by the rule,

$$x = \frac{445}{8} \pm \sqrt{-\frac{9801}{4} + \frac{198025}{64}},$$

or,
$$x = \frac{445}{8} \pm \frac{203}{8}:$$

taking the upper sign, $\quad x = \frac{648}{8} = 81$

" lower sign, $\quad x = \frac{242}{8} = \frac{121}{4}.$

8. Given, $\quad \frac{a}{b} - bx^4 + \frac{c}{d}x^2 = 0,\ $ to find $x$;

transposing and reducing.

$$x^4 - \frac{c}{bd}x^2 = \frac{a}{b^2}:$$

whence, $\quad x = \pm \sqrt{\frac{c}{2bd} \pm \sqrt{\frac{a}{b^2} + \frac{c^2}{4b^2d^2}}};$

reducing, $\quad x = \pm \sqrt{\frac{c \pm \sqrt{4ad^2 + c^2}}{2bd}}.$

### EXAMPLES OF REDUCTION OF EXPRESSIONS OF THE FORM OF
$$\sqrt{a \pm \sqrt{b}}.$$

4. Reduce to its simplest form, $\sqrt{28 + 10\sqrt{3}}.$

$$a = 28, \quad b = 300, \quad c = 22.$$

Applying the formula and considering only the upper sign,

$$\sqrt{28 + 10\sqrt{3}} = 5 + \sqrt{3}.$$

5. Reduce to its simplest form, $\sqrt{1 + 4\sqrt{-3}}.$

$$a = 1, \quad b = -48, \quad c = 7;$$

applying the formula, &c.,

$$\sqrt{1 + 4\sqrt{-3}} = 2 + \sqrt{-3}.$$

6. Reduce to its simplest form,

$$\sqrt{bc + 2b\sqrt{bc - b^2}} - \sqrt{bc - 2b\sqrt{bc - b^2}}.$$

$$a = bc, \quad b = 4b^2(bc - b^2), \quad c = b(c - 2b);$$

applying the formula to the 1st radical,

$$\sqrt{bc + 2b\sqrt{bc - b^2}} = \pm(\sqrt{bc - b^2} + b);$$

applying the formula to the 2d radical,

$$\sqrt{bc - 2b\sqrt{bc - b^2}} = \pm(\sqrt{bc - b^2} - b);$$

subtracting the second result from the first,

$$\sqrt{bc + 2b\sqrt{bc - b^2}} - \sqrt{bc - 2b\sqrt{bc - b^2}} = \pm 2b.$$

7. Reduce to its simplest form,

$$\sqrt{ab + 4c^2 - d^2 - 2\sqrt{4abc^2 - abd^2}}.$$

$$a = ab + 4c^2 - d^2, \quad b = 16abc^2 - 4abd^2, \quad c = ab - 4c^2 + d^2.$$

**181-183.]** EQUATIONS OF THE SECOND DEGREE. 97

Applying the formula,
$$\sqrt{ab + 4c^2 - d^2 - 2\sqrt{4abc^2 - abd^2}} = \pm(\sqrt{ab} - \sqrt{4c^2 - d^2}).$$

### SIMULTANEOUS EQUATIONS.

2. $\qquad \begin{cases} x + 2y = 7 & \dots \dots \dots (1) \\ x^2 + 3xy + y^2 = 31 & \dots \dots \dots (2) \end{cases}$

From (1),
$$x = 7 - 2y.$$
Substituting in (2),
$$49 - 28y + 4y^2 + 3y(7 - 2y) + y^2 = 31;$$
hence,
$$y = -\frac{7}{2} \pm \sqrt{\frac{49}{4} + \frac{18}{4}} = -\frac{7}{2} \pm \frac{11}{2};$$
$$\therefore\ y' = 2 \quad \text{and} \quad y'' = -9;$$
and
$$x' = 3 \quad \text{and} \quad x'' = 25.$$

3. $\qquad \begin{cases} 2x + y = 27 & \dots \dots \dots (1) \\ 3xy = 210 & \dots \dots \dots (2) \end{cases}$

From (1),
$$y = 27 - 2x.$$
Substituting in (2),
$$3x(27 - 2x) = 210;$$
hence,
$$x = \frac{27}{4} \pm \sqrt{\left(\frac{27}{4}\right)^2 - 35} = \frac{27 \pm 13}{4};$$
$$\therefore\ x' = 10 \quad \text{and} \quad x'' = 3\frac{1}{2}.$$
From (1)
$$y' = 7 \quad \text{and} \quad y'' = 20.$$

4. $\qquad \begin{cases} 2x - 3y - 1 = 0 & \dots \dots \dots (1) \\ 2x^2 + xy - 5y^2 = 20 & \dots \dots \dots (2) \end{cases}$

From (1),
$$x = \frac{3y + 1}{2};$$

from (2),
$$2\left(\frac{9y^2+6y+1}{4}\right)+\frac{3y^2+y}{2}-5y^2=20;$$
hence,
$$y=-\frac{7}{4}\pm\sqrt{\frac{49}{16}+\frac{312}{16}}=\frac{-7\pm 19}{4};$$
$$\therefore\ y'=3\quad\text{and}\quad y''=-6\tfrac{1}{2};$$
and from (5),
$$x'=5\quad\text{and}\quad x''=-9\tfrac{1}{4}.$$

5. $$\begin{cases}\dfrac{10x+y}{xy}=3 & \ldots\ldots\ldots\ (1)\\ y-x=2 & \ldots\ldots\ldots\ (2)\end{cases}$$

From (2),
$$y=2+x.$$
Clearing and substituting in (1),
$$10x+2+x=3x(2+x);$$
hence,
$$x=\frac{5}{6}\pm\sqrt{\frac{25}{36}+\frac{24}{36}}=\frac{5\pm 7}{6};$$
$$\therefore\ x'=2\quad\text{and}\quad x''=-\frac{1}{3}.$$
From (2),
$$y'=4\quad\text{and}\quad y''=\frac{5}{3}$$

7. $$\begin{cases}x^2-xy=6 & \ldots\ldots\ldots\ (1)\\ x^2+xy=66 & \ldots\ldots\ldots\ (2)\end{cases}$$

Adding (1) and (2) and dividing by 2,
$$x^2=36;\quad\therefore\ x'=6\quad\text{and}\quad x''=-6;$$
hence,
$$y'=5\quad\text{and}\quad y''=-5.$$

8. $$\begin{cases}x^2-xy=48 & \ldots\ldots\ldots\ (1)\\ xy-y^2=12 & \ldots\ldots\ldots\ (2)\end{cases}$$

Making
$$y = px,$$
$$x^2 - px^2 = 48; \quad \therefore \quad x^2 = \frac{48}{1-p} \quad \ldots \quad (3)$$

$$px^2 - p^2x^2 = 12; \quad \therefore \quad x^2 = \frac{12}{p-p^2} \quad \ldots \quad (4)$$

Equating (3) and (4) and dividing by 12,

$$\frac{4}{1-p} = \frac{1}{p-p^2}; \quad \therefore \quad 4p - 4p^2 = 1 - p;$$

hence,
$$p^2 - \frac{5}{4}p = -\frac{1}{4}; \quad \therefore \quad p = \frac{5}{8} \pm \sqrt{\frac{25}{64} - \frac{16}{64}},$$

or, $\quad p = 2 \quad$ and $\quad p = \frac{1}{4}.$

Using, $p = \frac{1}{4}$, we have from (3),

$x' = +8$ and $x'' = -8$; whence, $y' = +2$ and $y'' = -2$.

9. $\quad \begin{cases} x^2 + 4xy + 4y^2 = 256 & \ldots \ldots \quad (1) \\ 3y^2 - x^2 = 39 & \ldots \ldots \ldots \quad (2) \end{cases}$

Making
$$y = px,$$
$$x^2 + 4px^2 + 4p^2x^2 = 256 \quad \ldots \ldots \quad (3)$$
$$3p^2x^2 - x^2 = 39 \quad \ldots \ldots \ldots \quad (4)$$

From (3) and (4),
$$x^2 = \frac{256}{1 + 4p + 4p^2} \quad \ldots \ldots \ldots \quad (5)$$

$$x^2 = \frac{39}{3p^2 + 1} \quad \ldots \ldots \ldots \ldots \quad (6)$$

Equating,
$$\frac{256}{1 + 4p + 4p^2} = \frac{39}{3p^2 - 1};$$

hence,
$$768p^2 - 256 = 39 + 156p + 156p^2,$$

and
$$p^2 - \frac{13}{51}p = \frac{295}{612};$$

hence,
$$p = \frac{78 \pm 432}{612}, \quad \text{or} \quad p = \frac{5}{6} \quad \text{and} \quad p = -\frac{59}{102}.$$

Using the first value, we have from (6),
$$x^2 = 36; \quad \therefore \quad x' = 6 \quad \text{and} \quad x'' = -6;$$
and by substitution,
$$y' = 5 \quad \text{and} \quad y'' = -5.$$

Using the second value, we find,
$$x' = 102 \quad \text{and} \quad x'' = -102;$$
and by substitution,
$$y' = -59 \quad \text{and} \quad y'' = 59.$$

10.
$$\begin{cases} 6(x^2 + y^2) = 13xy & \cdots \cdots (1) \\ x^2 - y^2 = 20 & \cdots \cdots (2) \end{cases}$$

Making
$$y = px,$$
$$6(1 + p^2)x^2 = 13px^2 \cdots \cdots (3)$$
$$(1 - p^2)x^2 = 20 \cdots \cdots (4)$$

From (3), we have,
$$6 + 6p^2 = 13p.$$
Reducing,
$$p^2 - \frac{13}{6}p = -1;$$
$$\therefore p = \frac{13}{12} \pm \sqrt{\frac{169}{144} - \frac{144}{144}} = \frac{13 \pm 5}{12},$$
or
$$p = 1\tfrac{1}{2} \quad \text{and} \quad p = \frac{2}{3}.$$

From (4) we have,
$$x^2 = \frac{20}{1 - p^2} \cdots \cdots (5)$$

Using the second value of $p$, we have from (5),

$$x^2 = \frac{20}{1 - \frac{4}{9}} = 36, \quad \text{or} \quad x' = 6 \quad \text{and} \quad x'' = -6;$$

and by substitution, $y' = 4$ and $y'' = -4.$

## EQUATIONS OF A HIGHER DEGREE THAN THE FIRST, INVOLVING MORE THAN ONE UNKNOWN QUANTITY.

15. Given, $\begin{cases} x^2y + xy^2 = 6 \dots (1) \\ x^3y^2 + x^2y^3 = 12 \dots (2) \end{cases}$ to find $x$ and $y$.

Dividing (2) by (1), member by member,

$$xy = 2, \quad \text{or} \quad x = \frac{2}{y}.$$

Substituting this value of $x$ in (1) and reducing,

$$\frac{4}{y} + 2y = 6;$$

clearing of fractions and reducing,

$$y^2 - 3y = -2:$$

whence, $\quad y = \frac{3}{2} \pm \sqrt{-2 + \frac{9}{4}} = \frac{3}{2} \pm \frac{1}{2};$

or, $\quad y = 2, \quad \text{and} \quad y = 1;$

whence $\quad x = 1, \quad \text{and} \quad x = 2.$

16. Given, $\begin{cases} x^2 + x + y = 18 - y^2 \dots (1) \\ xy = 6 \dots \dots \dots \dots (2) \end{cases}$ to find $x$ and $y$.

Multiplying both members of (2) by 2 and adding the resulting equation to (1), member to member,

$$x^2 + 2xy + y^2 + x + y = 30,$$

or, $\quad (x + y)^2 + x + y = 30;$

whence, by the rule,

$$x + y = -\frac{1}{2} \pm \sqrt{30 + \frac{1}{4}} = -\frac{1}{2} \pm \frac{11}{2};$$

whence, $\quad x + y = 5, \quad \text{and} \quad x + y = -6.$

Taking the first value of $x + y$ and substituting in it, for $y$ its value $\frac{6}{x}$ derived from (2),

$$x + \frac{6}{x} = 5;$$

clearing of fractions and reducing,

$$x^2 - 5x = -6;$$

whence, $\quad x = \frac{5}{2} \pm \sqrt{-6 + \frac{25}{4}} = \frac{5}{2} \pm \frac{1}{2};$

or, $\quad x = 3, \quad \text{and} \quad x = 2; \quad \text{whence}, \quad y = 2, \quad \text{and} \quad y = 3.$

Taking the second value of $x + y$ and proceeding as before,

$$x + \frac{6}{x} = -6;$$

clearing of fractions, &c.,

$$x^2 + 6x = -6;$$

whence, $\quad x = -3 \pm \sqrt{-6 + 9} = -3 \pm \sqrt{3};$

and by substitution, $\quad y = -3 \mp \sqrt{3}.$

PROBLEMS GIVING RISE TO EQUATIONS OF A HIGHER DEGREE THAN THE FIRST CONTAINING MORE THAN ONE UNKNOWN QUANTITY.

2. To find four numbers, such that the sum of the first and fourth shall be equal to $2s$, the sum of the second and third equal to $2s'$, the sum of their squares equal to $4c^2$, and the product of the first and fourth equal to the product of the second and third.

Assuming the equations,

$$u + z = 2s \quad \cdots \quad (1)$$
$$x + y = 2s' \quad \cdots \quad (2)$$
$$u^2 + x^2 + y^2 + z^2 = 4c^2 \quad \cdots \quad (3)$$
$$uz = xy \quad \cdots \quad (4)$$

Multiplying both members of (4) by 2, and subtracting from (3), member from member,

$$u^2 - 2uz + z^2 + x^2 + 2xy + y^2 = 4c^2;$$

or, $\qquad (u - z)^2 + (x + y)^2 = 4c^2.$

Substituting for $x + y$ its value $2s'$ and transposing,

$$(u - z)^2 = 4c^2 - 4s'^2,$$

or, $\qquad u - z = \sqrt{4c^2 - 4s'^2}.$

Combining with (1),

$$u = s + \frac{\sqrt{4c^2 - 4s'^2}}{2} = s + \sqrt{c^2 - s'^2},$$

and, $\qquad z = s - \dfrac{\sqrt{4c^2 - 4s'^2}}{2} = s - \sqrt{c^2 - s'^2};$

reversing the order of the members of (4), and proceeding as before, we find in like manner,

$$x = s' + \sqrt{c^2 - s^2}$$

and, $\qquad y = s' - \sqrt{c^2 - s^2}.$

4. The sum of the squares of two numbers is expressed by $a$, and the difference of their squares by $b$: what are the numbers?

Let $x$ and $y$ denote the numbers.

From the conditions of the problem,

$$x^2 + y^2 = a \quad \cdots \quad (1)$$
$$x^2 - y^2 = b \quad \cdots \quad (2).$$

By adding, member to member,

$$2x^2 = a + b \quad \therefore \quad x = \pm\sqrt{\frac{a+b}{2}};$$

by subtracting,

$$2y^2 = a - b \quad \therefore \quad y = \pm\sqrt{\frac{a-b}{2}}.$$

5. What three numbers are they, which, multiplied two and two and each product divided by the third number, give the quotients, $a$, $b$, $c$?

Let $x$, $y$ and $z$, denote the numbers

From the conditions of the problem,

$$\frac{xy}{z} = a \quad \text{or,} \quad xy = az \quad \cdots \quad (1)$$

$$\frac{yz}{x} = b \quad \text{or,} \quad yz = bx \quad \cdots \quad (2)$$

$$\frac{xz}{y} = c \quad \text{or,} \quad xz = cy \quad \cdots \quad (3).$$

Multiplying (1), (2) and (3) together, member by member,

$$x^2y^2z^2 = abcxyz;$$

dividing both members by $xyz$,

$$xyz = abc \quad \cdots \quad (4);$$

substituting in (4) the value of $xy$ taken from (1), and dividing both members by $a$,

$$z^2 = bc \quad \therefore \quad z = \sqrt{bc}.$$

Substituting the value of $yz$ and dividing by $b$,

$$x^2 = ac \quad \therefore \quad x = \sqrt{ac}.$$

Substituting the value of $xz$ and dividing both members by $c$,

$$y^2 = ab \quad \therefore \quad y = \sqrt{ab}.$$

6. The sum of two numbers is 8, and the sum of their cubes is 152; what are the numbers?

Let $x$ and $y$ denote the numbers.

From the conditions,

$$x + y = 8 \quad \cdots \cdots (1)$$
$$x^3 + y^3 = 152 \quad \cdots \cdots (2);$$

cubing both members of (1),

$$x^3 + 3x^2y + 3xy^2 + y^3 = 512 \quad \cdots \cdots (3);$$

subtracting (2) from (3), member from member, and dividing both members by 3,

$$x^2y + xy^2 = 120 \quad \cdots \cdots (4);$$

substituting the value of $x$ taken from (1),

$$(64 - 16y + y^2)\,y + (8 - y)\,y^2 = 120,$$

or, reducing, $\quad y^2 - 8y = -15;$

whence, $\quad y = 4 \pm \sqrt{-15 + 16} = 4 \pm 1 \quad \therefore \quad y = 5, \quad y = 3;$

whence, from (1) $\qquad\qquad\qquad\qquad x = 3, \quad x = 5.$

7. Find two numbers, whose difference added to the difference of their squares is 150, and whose sum added to the sum of their squares, is 330.

Let $x$ and $y$ denote the numbers.

From the conditions of the problem,

$$x^2 - y^2 + x - y = 150 \quad \cdots \quad (1),$$
$$x^2 + y^2 + x + y = 330 \quad \cdots \quad (2);$$

adding member to member, and reducing,
$$x^2 + x = 240;$$

whence, $\quad x = -\dfrac{1}{2} \pm \sqrt{240 + \dfrac{1}{4}} = -\dfrac{1}{2} \pm \dfrac{31}{2};$

or, considering only the positive solution,
$$x = 15;$$

whence, from (1), by substitution,
$$y = 9.$$

8. There are two numbers whose difference is 15, and half their product is equal to the cube of the lesser number: what are the numbers?

Let $x$ and $y$ denote the numbers.

From the conditions of the problem,
$$x - y = 15 \quad \cdots \quad (1),$$
$$\dfrac{xy}{2} = y^3, \quad \text{or} \quad x = 2y^2 \quad \cdots \quad (2);$$

substituting in (1) and dividing both members by 2,
$$y^2 - \dfrac{y}{2} = \dfrac{15}{2};$$

whence,
$$y = \dfrac{1}{4} \pm \sqrt{\dfrac{15}{2} + \dfrac{1}{16}} = \dfrac{1}{4} \pm \dfrac{11}{4}:$$

considering only the positive solution,
$$y = 3; \quad \text{whence, from (1)}, \quad x = 18.$$

9. What two numbers are those whose sum multiplied by the greater, is equal to 77; and whose difference, multiplied by the lesser, is equal to 12?

Let $x$ and $y$ denote the numbers.

From the conditions,

$$(x + y) x = 77, \quad \text{or} \quad x^2 + xy = 77 \quad \cdots \quad (1);$$

$$(x - y) y = 12, \quad \text{or} \quad xy - y^2 = 12 \quad \cdots \quad (2);$$

make $x = py$; whence,

$$(p^2 + p) y^2 = 77, \quad \text{or,} \quad y^2 = \frac{77}{p^2 + p} \quad \cdots \quad (3),$$

$$(p - 1) y^2 = 12, \quad \text{or,} \quad y^2 = \frac{12}{p - 1} \quad \cdots \quad (4);$$

equating the second members and reducing,

$$p^2 - \frac{65}{12} p = -\frac{77}{12};$$

whence, $\quad p = \frac{65}{24} \pm \sqrt{-\frac{77}{12} + \frac{4225}{576}} = \frac{65}{24} \pm \frac{23}{24};$

taking the upper sign,

$$p = \frac{88}{24} = \frac{11}{3};$$

substituting in (4), $\quad y = \sqrt{\frac{36}{8}} = \frac{3}{2} \sqrt{2};$

whence, $\quad x = \frac{11}{2} \sqrt{2};$

taking the lower sign, $\quad p = \frac{42}{24} = \frac{21}{12};$

substituting in (4),
$$y = \sqrt{\frac{144}{9}} = 4; \quad \text{whence,} \quad x = 7.$$

10. Divide 100 into two such parts, that the sum of their square roots may be 14.

Let $x$ and $y$ denote the parts.

From the conditions,
$$x + y = 100 \quad \cdots \quad (1),$$
$$\sqrt{x} + \sqrt{y} = 14 \quad \cdots \quad (2):$$

squaring both members of (2) and subtracting (1), member from member.
$$2\sqrt{xy} = 96, \quad \text{or} \quad \sqrt{xy} = 48, \quad \text{or} \quad xy = 2304:$$

substituting for $y$ its value, $100 - x$,
$$100x - x^2 = 2304, \quad \text{or} \quad x^2 - 100x = -2304.$$

whence, by the rule,
$$x = 50 \pm \sqrt{196} = 50 \pm 14;$$
hence, $\quad x = 64, \quad x = 36,$
and $\quad y = 36, \quad y = 64.$

11. It is required to divide the number 24 into two such parts, that their product may be equal to 35 times their difference.

Let $x$ and $y$ denote the parts.

From the conditions of the problem,
$$x + y = 24 \quad \cdots \quad (1)$$
$$xy = 35(x - y) \quad \cdots \quad (2);$$

substituting in (2) the value, $y = 24 - x$,
$$24x - x^2 = 35(2x - 24), \quad \text{or,} \quad 24x - x^2 = 70x - 840;$$

whence, $\quad x^2 + 46x = 840$;

by the rule,
$$x = -23 \pm \sqrt{840 + 529} = -23 \pm 37;$$

ence, taking the upper sign, $x = 14$;

by substitution, in (1), $\quad y = 10$.

12. What two numbers are those, whose product is 255, and the sum of whose square is 514?

Let $x$ and $y$ denote the numbers.

From the conditions, $\quad xy = 255 \quad \cdots \quad (1)$

$$x^2 + y^2 = 514 \quad . \quad . \quad (2),$$

multiplying both members of (1) by 2, adding and subtracting the resulting equation to and from (2), member by member,

$$x^2 + 2xy + y^2 = 1024 \quad . \quad . \quad (3)$$
$$x^2 - 2xy + y^2 = 4 \quad . \quad . \quad . \quad (4);$$

extracting the square root of both members,

$$x + y = 32,$$
$$x - y = 2,$$

whence, $\quad x = 17; \quad y = 15$.

13. There is a number expressed by two digits, which, when divided by the sum of the digits, gives a quotient greater by 2 than the first digit; but if the digits be inverted, and the resulting number be divided by a number greater by 1 than the sum of the digits, the quotient will exceed the former quotient by 2: what is the number?

Let $x$ and $y$ denote the digits; then will

$$10x + y \text{ denote the number.}$$

From the conditions,

$$\frac{10x + y}{x + y} = x + 2 \quad \ldots \quad (1)$$

$$\frac{10y + x}{x + y + 1} = x + 4 \quad \ldots \quad (2);$$

clearing of fractions, and reducing,

$$8x - y - x^2 - xy = 0 \quad \cdots \quad (3),$$
$$6y - 4x - x^2 - xy = 4 \quad \cdots \quad (4);$$

by subtraction,

$$7y - 12x = 4; \quad \therefore \quad y = \frac{12x + 4}{7};$$

substituting in (3),

$$8x - \frac{12x + 4}{7} - x^2 - \frac{12x^2 + 4x}{7} = 0,$$

clearing of fractions and reducing,

$$x^2 - \frac{40}{19}x = -\frac{4}{19}, \quad \text{whence,}$$

$$x = \frac{20}{19} \pm \sqrt{-\frac{4}{19} + \frac{400}{361}} = \frac{20}{19} \pm \frac{18}{19}.$$

Taking the upper sign, gives $x = 2$; whence, $y = 4$.

14. A regiment, in garrison, consisting of a certain number of companies, receives orders to send 216 men on duty, each company to furnish an equal number. Before the order was executed, three of the companies were sent on another service, and it was then found that each company that remained would have to send 12 men additional, in order to make up the complement, 216. How many

companies were in the regiment, and what number of men did each of the remaining companies send?

Let $x$ denote the number of companies, and
$y$ " " " each should send; then,
$y + 12$ will denote the number sent by each.

From the conditions of the problem,

$$xy = 216 \quad \cdot \quad \cdot \quad \cdot \quad (1),$$
$$(x - 3)(y + 12) = 216 \quad \cdot \quad \cdot \quad \cdot \quad (2);$$

Performing operations, subtracting and reducing,

$$4x - y = 12. \quad \therefore \quad y = 4x - 12;$$

substituting in (1), $4x^2 - 12x = 216$, or $x^2 - 3x = 54$;

whence, $$x = \frac{3}{2} \pm \sqrt{54 + \frac{9}{4}} = \frac{3}{2} \pm \frac{15}{2};$$

taking the upper sign,

$$x = 9; \quad \text{hence,} \quad y = 24, \quad \text{and} \quad y + 12 = 36.$$

15. Find three numbers such, that their sum shall be 14, the sum of their squares equal to 84, and the product of the first and third equal to the square of the second.

Let $x$, $y$ and $z$ denote the numbers.

From the conditions of the problem,

$$x + y + z = 14 \quad \cdot \quad \cdot \quad \cdot \quad (1),$$
$$x^2 + y^2 + z^2 = 84 \quad \cdot \quad \cdot \quad \cdot \quad (2),$$
$$xz = y^2 \quad \cdot \quad \cdot \quad \cdot \quad (3).$$

Multiplying both members of (3), by 2, adding to (2) and reducing,

$$x^2 + 2xz + z^2 = 84 + y^2 \; ; \quad \therefore \; x + z = \sqrt{84 + y^2} \cdot \cdot \; (4)$$

from (1), $\qquad\qquad\qquad\qquad x + z = 14 - y \; \cdot \cdot \cdot \cdot \; (5)\; ;$

equating the second members of (4) and (5) and squaring,

$$84 + y^2 = 196 - 28y + y^2 \; ; \quad \therefore \; y = 4.$$

Substituting in (1) and (3),

$$x + z = 10 \; \cdot \; \cdot \; \cdot \; \cdot \; (6)$$
$$xz = 16 \; \cdot \; \therefore \; x = \frac{16}{z}.$$

Substituting in (6) and reducing,

$$z^2 - 10z = -16$$
$$\therefore \quad z = 5 \pm \sqrt{9} = 5 \pm 3,$$
$$z = 8 \; ; \quad z = 2,$$

and by substitution, $\quad x = 2 \; ; \quad x = 8.$

16. It is required to find a number, expressed by three digits, such, that the sum of the squares of the digits shall be 104; the square of the middle digit to exceed twice the product of the other two by 4; and if 594 be subtracted from the number, the remainder will be expressed by the same figures, but with the extreme digits reversed.

Let $x$, $y$ and $z$ denote the digits;

then, $\qquad 100x + 10y + z \quad$ will denote the number.

From the conditions of the problem,

$$x^2 + y^2 + z^2 = 104 \; \cdot \; \cdot \; \cdot \; (1)$$
$$y^2 - 2xz = 4 \quad \cdot \; \cdot \; \cdot \; (2)$$
$$100x + 10y + z - 594 = 100z + 10y + x \quad (3) \; ;$$

subtracting (2) from (1), member from member,

$$x^2 + 2xz + z^2 = 100 \quad \therefore \quad x + z = 10;$$

reducing (3) $\qquad\qquad\qquad\qquad x - z = 6;$

hence, $\qquad\qquad x = 8, \text{ and } z = 2.$

By substitution, $\quad y = 6,$ and the number is 862.

17. A person has three kinds of goods which together cost $230$\frac{5}{24}$. A pound of each article costs as many $\frac{1}{24}$ dollars as there are pounds in that article: he has one-third more of the second than of the first, and $3\frac{1}{2}$ times as much of the third as of the second. How many pounds has he of each article?

Let $x$, $y$ and $z$ denote the number of pounds of each article.

From the conditions of the problem,

$$\frac{x^2}{24} + \frac{y^2}{24} + \frac{z^2}{24} = \frac{5525}{24}, \quad \text{or,} \quad x^2 + y^2 + z^2 = 5525 \quad (1)$$

$$y = \frac{4}{3}x \qquad \therefore \qquad y^2 = \frac{16}{9}x^2 \quad \ldots \quad (2)$$

$$z = \frac{7}{2}y \quad \therefore \quad z = \frac{14}{3}x, \quad \text{and,} \quad z^2 = \frac{196}{9}x^2 \quad \ldots \quad (3);$$

substituting in (1) and reducing,

$$x^2 = 225 \quad \therefore \quad x = 15,$$

substituting in (2) and (3),

$$y = 20; \qquad z = 70.$$

18. Two merchants each sold the same kind of stuff: the second sold 3 yards more of it than the first, and together, they received 35 dollars. The first said to the second, "I would have received 24 dollars for your stuff." The other replied, "And I would have

received 12½ dollars for yours." How many yards did each of them sell?

Let $x$ and $y$ denote the number of yards sold by each.

Then will $\dfrac{24}{y}$ denote the price the first received per yard,

and $\dfrac{25}{2x}$ will denote the price the second received per yard.

From the conditions of the problem,

$$x + 3 = y,$$

$$\frac{24x}{y} + \frac{25y}{2x} = 35, \quad \text{or,} \quad 48x^2 + 25y^2 = 70xy:$$

substituting in the second equation the value of $y$ taken from the first,

$$48x^2 + 25(x^2 + 6x + 9) = 70x^2 + 210x;$$

reducing, $\qquad x^2 - 20x = -75;$

whence, $\qquad x = 10 \pm \sqrt{25} = 10 \pm 5,$

or, $\qquad x = 15; \qquad x = 5;$

substituting, $\quad y = 18; \qquad y = 8.$

19. A widow possessed 13000 dollars, which she divided into two parts, and placed them at interest, in such a manner, that the incomes from them were equal. If she had put out the first portion at the same rate as the second, she would have drawn for this part 360 dollars interest; and if she had placed the second out at the same rate as the first, she would have drawn for it 490 dollars interest. What were the two rates of interest?

. Let $x$ and $y$ denote the rates per cent.

Let $z$ denote the 1st portion; then will $13000 - z$ denote the 2d.

From the conditions of the problem,

$$\frac{xz}{100} = \frac{(13000-z)y}{100}, \quad \text{or,} \quad xz = 13000y - zy \quad \cdots \quad (1),$$

$$\frac{zy}{100} = 360, \quad \text{or,} \quad zy = 36000 \quad \cdots \quad (2),$$

$$\frac{(13000-z)x}{100} = 490, \quad \text{or,} \quad 13000x - zx = 49000 \quad \cdots \quad (3).$$

Substituting in (1) the values of $zy$ and $zx$ taken from (2) and (3) and reducing, we find, $x = y + 1$.

Substituting this value of $x$ and the value of $z$ taken from (2) in (1), and reducing, we find

$$y^2 - \frac{72}{13}y = \frac{36}{13};$$

whence, $y = \dfrac{36}{13} \pm \sqrt{\dfrac{36}{13} + \dfrac{1296}{169}} = \dfrac{36 \pm 42}{13}; \quad \therefore y = 6;$

by substitution,

$$x = 7, \quad \text{and} \quad z = 6000, \quad 13000 - z = 7000.$$

## ADDITION AND SUBTRACTION OF RADICALS.

1. $\sqrt{48ab^2} = 4b\sqrt{3a}$, and $b\sqrt{75a} = 5b\sqrt{3a}$; $\therefore$ Ans. $9b\sqrt{3a}$.

2. $3\sqrt[3]{4a^2} = 3\sqrt[3]{2a}$; $2\sqrt[3]{2a}$, $\therefore$ Ans. $5\sqrt[3]{2a}$.

3. $2\sqrt{45} = 6\sqrt{5}$; $3\sqrt{5}$, $\therefore$ Ans. $9\sqrt{5}$.

4. $3a\sqrt[4]{b} - 2c\sqrt[4]{b}$, $\therefore$ Ans. $(3a - 2c)\sqrt[4]{b}$.

5. $3\sqrt[3]{4a^2} - 2\sqrt[3]{2a}$;

   $3\sqrt[3]{4a^2} = 3\sqrt[3]{2a}$; $\therefore$ Ans. $\sqrt[3]{2a}$.

6. $\sqrt{243} = 9\sqrt{3}$; $\sqrt{27} = 3\sqrt{3}$; $\sqrt{48} = 4\sqrt{3}$. Ans. $16\sqrt{3}$.

7. $2\sqrt{8a^3} = 4a\sqrt{2a}$; $\quad -7a\sqrt{18a} = -21a\sqrt{2a}$;

$5\sqrt{72a^3} = 30a\sqrt{2a}$; $\quad -\sqrt{50ab^2} = -5b\sqrt{2a}$;

$\therefore$ Ans. $(13a - 5b)\sqrt{2a}$.

8. $12\sqrt[3]{\tfrac{1}{4}} = 6\sqrt[3]{2}$; $\quad 3\sqrt[3]{\tfrac{1}{32}} = \tfrac{3}{4}\sqrt[3]{2}$;

$\therefore$ Ans. $\tfrac{27}{4}\sqrt[3]{2}$.

9. $\sqrt[3]{8a^3b + 16a^4} = 2a\sqrt[3]{b + 2a}$, and

$\sqrt[3]{b^4 + 2ab^3} = b\sqrt[3]{b + 2a}$, $\therefore$ Ans. $(2a - b)\sqrt[3]{b + 2a}$.

10. $3\sqrt[6]{4a^2} = 3\sqrt[3]{2a}$; $\quad 2\sqrt[3]{2a}$; $\quad \therefore$ Ans. $\sqrt[3]{2a}$.

## MULTIPLICATION OF RADICALS.

5. $\sqrt{2} \times \sqrt[3]{3} = \sqrt[12]{64} \times \sqrt[12]{81} = \sqrt[12]{5184}$

$\sqrt[4]{\tfrac{1}{2}} \times \sqrt[3]{\tfrac{1}{3}} = \sqrt[12]{\tfrac{1}{8}} \times \sqrt[12]{\tfrac{1}{81}} = \sqrt[12]{\tfrac{1}{648}}$; $\therefore$ Ans. $\sqrt[12]{8}$.

6. $2\sqrt{15} = 2\sqrt[6]{3375}$; $\quad 3\sqrt[3]{10} = 3\sqrt[6]{100}$; $\therefore$ Ans. $6\sqrt[6]{337500}$

7. $4\sqrt[5]{\tfrac{2}{3}} = 4\sqrt[10]{\tfrac{4}{9}}$; $\quad 2\sqrt{\tfrac{3}{4}} = 2\sqrt[10]{\tfrac{243}{1024}}$; $\therefore$ Ans. $8\sqrt[10]{\tfrac{27}{256}}$.

8. $\sqrt{2} = \sqrt[12]{64}$; $\quad \sqrt[3]{3} = \sqrt[12]{81}$; $\quad \sqrt[4]{5} = \sqrt[12]{125}$;

$\therefore$ Ans. $\sqrt[12]{648000}$.

9. $\sqrt[7]{\tfrac{4}{3}} = \sqrt[42]{\tfrac{4096}{729}}$; $\quad \sqrt[3]{\tfrac{1}{2}} = \sqrt[42]{\tfrac{1}{16384}}$; $\quad \sqrt[14]{6} = \sqrt[42]{216}$;

$\therefore$ Ans. $\sqrt[42]{\tfrac{2}{27}}$

10. The product by the rule for multiplication is

$\tfrac{28}{3} + 8\sqrt{\tfrac{7}{6}} + 5\sqrt{\tfrac{7}{6}} + \tfrac{10}{2} = \tfrac{43}{3} + 13\sqrt{\tfrac{7}{6}} = \tfrac{43}{3} + \tfrac{13}{6}\sqrt{42}\ldots$

11. $-\sqrt{\dfrac{9m^2}{a^2+b^2}} = -\dfrac{3m}{\sqrt{a^2+b^2}}$; $-\dfrac{a^2+b^2}{3m} = -\dfrac{\sqrt{a^2+b^2}\sqrt{a^2+b^2}}{3m}$

$\therefore$ Ans. $= \sqrt{a^2+b^2}$.

12. The product equals

$$(\sqrt{x})^2 - (\sqrt{y})^2 = x - y. \quad Ans.$$

## DIVISION OF RADICALS.

2. $2\sqrt{3} \times \sqrt[3]{4} = 2\sqrt[12]{729} \times \sqrt[12]{256} = 2\sqrt[12]{729 \times 256}$,

$\frac{1}{2}\sqrt[4]{2} \times \sqrt[3]{3} = \frac{1}{2}\sqrt[12]{8} \times \sqrt[12]{81} = \frac{1}{2}\sqrt[12]{8 \times 81}$

$2\sqrt[12]{729} \times 256 \div \frac{1}{2}\sqrt[12]{8 \times 81} = 4\sqrt[12]{\dfrac{729 \times 256}{8 \times 81}} = 4\sqrt[12]{288}.$ Ans.

3. $\quad \dfrac{1}{2}\sqrt{\dfrac{1}{2}} = \dfrac{1}{4}\sqrt{2};\quad \sqrt{2} + 3\sqrt{\dfrac{1}{2}} = \dfrac{5}{2}\sqrt{2};$

$\dfrac{1}{4}\sqrt{2} \div \dfrac{5}{2}\sqrt{2} = \dfrac{1}{10}.\quad$ Ans.

4. $\dfrac{1}{\sqrt[4]{a} + \sqrt[4]{b}}$: multiplying both terms by $\sqrt[4]{a} - \sqrt[4]{b}$, we have

$$\dfrac{\sqrt[4]{a} - \sqrt[4]{b}}{\sqrt{a} - \sqrt{b}};$$

again multiplying both terms by $\sqrt{a} + \sqrt{b}$, we finally obtain

$$\dfrac{\sqrt[4]{a^3} - \sqrt[4]{a^2b} + \sqrt[4]{ab^2} - \sqrt[4]{b^3}}{a - b} \quad Ans.$$

5. $\quad \sqrt[4]{a} + \sqrt[4]{b} \div \sqrt[4]{a} - \sqrt[4]{b}$:

multiplying both terms by $\sqrt[4]{a} + \sqrt[4]{b}$, we have

$$\sqrt{a} + 2\sqrt[4]{ab} + \sqrt{b}, \text{ and } \sqrt{a} - \sqrt{b};$$

multiplying both by $\sqrt{a} + \sqrt{b}$, and we have

$$\dfrac{a + b + 2\sqrt{ab} + 2\sqrt[4]{a^3b} + 2\sqrt[4]{ab^3}}{a - b}. \quad Ans.$$

6. $\frac{43}{3} + \frac{13}{6}\sqrt{42} = \frac{1}{6}(86 + 13\sqrt{42})$;

$\sqrt{\frac{7}{3}} + 2\sqrt{\frac{1}{2}} = \frac{1}{6}(2\sqrt{21} + 6\sqrt{2})$; $\therefore \frac{86 + 13\sqrt{42}}{2\sqrt{21} + 6\sqrt{2}}$ (1)

Multiplying both terms of (1) by $2\sqrt{21} - 6\sqrt{2}$,

$$\frac{172\sqrt{21} + 26\sqrt{21 \times 42} - 516\sqrt{2} - 78\sqrt{84}}{84 - 72} \quad \cdot \cdot \text{ (2)}$$

$26\sqrt{21 \times 42} = 546\sqrt{2}$; $\quad -78\sqrt{84} = -156\sqrt{21}$;

$\therefore \quad \frac{16}{12}\sqrt{21} + \frac{30}{12}\sqrt{2} = 4\sqrt{\frac{7}{3}} + 5\sqrt{\frac{1}{2}}$.

## FRACTIONAL EXPONENTS.

1. $a^{\frac{3}{4}}b^{-\frac{1}{2}}c^{-1}$ multiplied by $a^2 b^{\frac{2}{3}} c^{\frac{3}{5}}$

$\frac{3}{4} + 2 = \frac{11}{4}$; $\quad -\frac{1}{2} + \frac{2}{3} = +\frac{1}{6}$; $\quad -1 + \frac{3}{5} = -\frac{2}{5}$;

hence, $\qquad a^{\frac{3}{4}}b^{-\frac{1}{2}}c^{-1} \times a^2 b^{\frac{2}{3}} c^{\frac{3}{5}} = a^{\frac{11}{4}} b^{\frac{1}{6}} c^{-\frac{2}{5}}$.

2. $3a^{-2}b^{\frac{2}{3}}$ multiplied by $2a^{-\frac{4}{5}}b^{\frac{1}{2}}c^2$.

$-2 - \frac{4}{5} = -\frac{14}{5}$; $\quad \frac{2}{3} + \frac{1}{2} = \frac{7}{6}$; $\quad 0 + 2 = 2$;

hence, $\qquad 3a^{-2}b^{\frac{2}{3}} \times 2a^{-\frac{4}{5}}b^{\frac{1}{2}}c^2 = 6a^{-\frac{14}{5}}b^{\frac{7}{6}}c^2$.

3. $6a^{-\frac{1}{2}}b^4 c^{-m}$ multiplied by $5a^{\frac{1}{3}}b^{-5}c^n$.

$-\frac{1}{2} + \frac{1}{3} = -\frac{1}{6}$; $\quad 4 - 5 = -1$; $\quad -m + n = n - m$;

hence, $\qquad 6a^{-\frac{1}{2}}b^4 c^{-m} \times 5a^{\frac{1}{3}}b^{-5}c^n = 3a^{-\frac{1}{6}}b^{-1}c^{n-m}$.

4. $\left(\frac{2}{3}a^{\frac{1}{3}}\right)$ multiplied by $\frac{2}{3}a^{\frac{1}{3}}$.

$\frac{2}{3} \times \frac{2}{3} = \frac{4}{9}$; $\quad \frac{1}{3} \times 2 = \frac{2}{3}$;

hence, $\qquad \left(\frac{2}{3}a^{\frac{1}{3}}\right)^2 = \frac{4}{9}a^{\frac{2}{3}}$.

5. $\left(\frac{1}{3}a^{\frac{1}{2}}\right)^3 = \left(\frac{1}{3}a^{\frac{1}{2}}\right) \times \frac{1}{3}a^{\frac{1}{2}} \times \frac{1}{3}a^{\frac{1}{2}}$.

$$\frac{1}{3} \times \frac{1}{3} \times \frac{1}{3} = \frac{1}{27}; \quad \frac{1}{2} + \frac{1}{2} + \frac{1}{2} = \frac{3}{2};$$

hence, $\left(\frac{1}{3}a^{\frac{1}{2}}\right)^3 = \frac{1}{27}a^{\frac{3}{2}}$.

6. $a^{\frac{2}{3}}$ divided by $a^{-\frac{3}{4}}$.

$$\frac{2}{3} - \left(-\frac{3}{4}\right) = \frac{2}{3} + \frac{3}{4} = \frac{17}{12};$$

hence, $a^{\frac{2}{3}} \div a^{-\frac{3}{4}} = a^{\frac{17}{12}}$.

7. $a^{\frac{3}{4}}$ divided by $a^{\frac{4}{5}}$.

$$\frac{3}{4} - \frac{4}{5} = -\frac{1}{120};$$

hence, $a^{\frac{3}{4}} \div a^{\frac{4}{5}} = a^{-\frac{1}{20}}$.

8. $a^{\frac{2}{5}} \times b^{\frac{3}{4}}$ divided by $a^{-\frac{1}{2}}b^{\frac{7}{8}}$.

$$\frac{2}{5} - \left(\frac{1}{2}\right) = \frac{2}{5} + \frac{1}{2} = \frac{9}{10}; \quad \frac{3}{4} - \frac{7}{8} = -\frac{1}{8};$$

hence, $a^{\frac{2}{5}}b^{\frac{3}{4}} \div a^{-\frac{1}{2}}b^{\frac{7}{8}} = a^{\frac{9}{10}}b^{-\frac{1}{8}}$.

9. $32a^{\frac{1}{2}}b^6c^{\frac{5}{2}}$ divided by $8a^{\frac{1}{6}}b^5c^{-\frac{3}{2}}$.

$$32 \div 8 = 4; \quad \frac{1}{2} - \frac{1}{6} = \frac{1}{3}; \quad 6 - 5 = 1; \quad \frac{5}{2} - \left(\frac{3}{2}\right) = 4;$$

hence, $32a^{\frac{1}{2}}b^6c^{\frac{5}{2}} \div 8a^{\frac{1}{6}}b^5c^{-\frac{3}{2}} = 4a^{\frac{1}{3}}bc^4$.

10. $64a^9b^{\frac{7}{2}}c^{-\frac{3}{5}}$ divided by $32a^{-9}b^{-\frac{3}{2}}c^{-\frac{3}{5}}$.

$64 \div 32 = 2$; $9 - (-9) = 18$; $\frac{7}{2} - \left(-\frac{3}{2}\right) = 5$; $-\frac{3}{5} - \left(-\frac{3}{5}\right) = 0$;

hence, $64a^9b^{\frac{7}{2}}c^{-\frac{3}{5}} \div 32a^{-9}b^{-\frac{3}{2}}c^{-\frac{3}{5}} = 4a^{18}b^5$.

11. $\sqrt[3]{a^{\frac{2}{3}}} \times \sqrt[4]{b^{\frac{4}{3}}} \times \sqrt{c^{\frac{5}{4}}}$.

$\sqrt[3]{a^{\frac{2}{3}}} = a^{\frac{2}{9}}$; $\sqrt[4]{b^{\frac{4}{3}}} = b^{\frac{4}{12}} = b^{\frac{1}{3}}$; $\sqrt{c^{\frac{5}{4}}} = c^{\frac{5}{8}} = c^{\frac{5}{8}}$;

hence, $\sqrt[3]{a^{\frac{2}{3}}} \times \sqrt[4]{b^{\frac{4}{3}}} \times \sqrt{c^{\frac{5}{4}}} = a^{\frac{2}{9}}b^{\frac{1}{3}}c^{\frac{5}{8}}$

12. Reduce $\dfrac{2\sqrt{2} \times (3)^{\frac{1}{3}}}{\frac{1}{3}\sqrt{2}}$ to its simplest terms.

Cancelling the $\sqrt{2}$, and writing the quotient of $2 \div \frac{1}{2}$, we have

$$4\sqrt[3]{3}. \quad Ans.$$

13. Reduce $\left\{ \dfrac{\frac{1}{2}(2)^{\frac{1}{3}}\sqrt[3]{3}}{2\sqrt[4]{2}\,(3)^{\frac{1}{2}}} \right\}^4$ to its simplest terms.

Raising both terms to the 4th power, we have

$$\frac{(\tfrac{1}{2})^4\, 2^{\frac{4}{3}}\, 3^{\frac{4}{3}}}{(2)^4\, 2\, (3)^2} = \frac{(\tfrac{1}{2})^4\, (3)\sqrt[3]{3}}{(2)^3\, (3)^2} = \frac{\sqrt[3]{3}}{(2)^7 \times 3} = \frac{1}{384}\sqrt[3]{3}.$$

14. Reduce $\sqrt{\left\{ \dfrac{(\tfrac{1}{2})^3 + \sqrt{3\tfrac{1}{2}}}{2\sqrt{2}\cdot(\tfrac{3}{4})^{\frac{1}{2}}} \right\}^{\frac{1}{2}}}$ to its simplest terms.

Since the square root of the square root is equal to the 4th root, we need only operate on the terms of the fraction:

$$\frac{(\tfrac{1}{2})^3 + \sqrt{3\tfrac{1}{2}}}{2\sqrt{2}\times(3)^{\frac{1}{2}}} = \frac{\tfrac{1}{8} + \sqrt{\tfrac{7}{2}}}{2\sqrt{2}\times\tfrac{1}{2}\sqrt{3}} = \frac{\tfrac{1}{8} + \sqrt{\tfrac{7}{2}}}{\sqrt{6}}.$$

Multiplying both terms of the fraction by the $\sqrt{6}$, we have

$$\frac{\tfrac{1}{6}(\tfrac{1}{8}\sqrt{6} + \sqrt{\tfrac{42}{2}}) = \tfrac{1}{6}(\tfrac{1}{8}\sqrt{6} + \sqrt{21})}{6} \; : \; \text{hence,}$$

$$\sqrt[4]{\tfrac{1}{6}(\tfrac{1}{8}\sqrt{6} + \sqrt{21})}.$$

15. Multiply $a^{\frac{3}{2}} + a^2 b^{\frac{1}{2}} + a^{\frac{3}{2}} b^{\frac{2}{3}} + ab + a^{\frac{1}{2}} b^{\frac{4}{3}} + b^{\frac{5}{3}}$,

$$\frac{a^{\frac{1}{2}} - b^{\frac{1}{3}}\,;}{a^3 + a^{\frac{5}{2}} b^{\frac{1}{3}} + a^2 b^{\frac{2}{3}} + a^{\frac{3}{2}} b + ab^{\frac{4}{3}} + a^{\frac{1}{2}} b^{\frac{5}{3}}}$$

$$\frac{- a^{\frac{5}{2}} b^{\frac{1}{3}} - a^2 b^{\frac{2}{3}} - a^{\frac{3}{2}} b - ab^{\frac{4}{3}} - a^{\frac{1}{2}} b^{\frac{5}{3}} - b^2}{a^3 - b^2. \quad Ans.}$$

16. Divide $a^{7/3} - a^2 b^{-2/3} - a^{1/3} b + b^{1/3}$ | $a^{1/3} - b^{-2/3}$
$\underline{a^{7/3} - a^2 b^{-2/3}}$ | $a^2 - b$
$\qquad - a^{1/3} b + b^{1/3}$
$\qquad \underline{- a^{1/3} b + b^{1/3}}.$

17. $x^{3/4} - \tfrac{3}{4} a^{3/4}$ | $x^{1/2} - \tfrac{1}{2} a^{1/3}$
 $\underline{x^{3/4} - \tfrac{1}{2} a^{1/2} x^{1/4}}$ | $x^{1/4} + \tfrac{1}{2} a^{1/2} x^{-1/4} + \tfrac{1}{4} a x^{-3/4} + $ &c.
 $\qquad \tfrac{1}{2} a^{1/2} x^{1/4} - \tfrac{3}{4} a^{3/4}$
 $\qquad \underline{\tfrac{1}{2} a^{1/2} x^{1/4} - \tfrac{1}{4} a x^{-1/4}}$
 $\qquad\qquad + \tfrac{1}{4} a x^{-1/4} - \tfrac{3}{4} a^{3/4}$
 $\qquad\qquad\quad$ &c., &c.

18. $x^{-1} + x^{-1/2} y^{-1/2} + y^{-1}$
$\underline{y \ + x^{1/2} y^{1/2} \ \ + x}$
$y x^{-1} + x^{-1/2} y^{1/2} + 1$
$\qquad x^{-1/2} y^{1/2} + 1 + x^{1/2} y^{-1/2}$
$\qquad\qquad + 1 + x^{1/2} y^{-1/2} + x y^{-1}$
$\overline{y x^{-1} + 2 x^{-1/2} y^{1/2} + 3 + 2 x^{1/2} y^{-1/2} + x y^{-1}}$

19. $\left(\dfrac{a^{2/3}}{2} - \dfrac{b^{2/3}}{3}\right)^2 = \dfrac{a^{4/3}}{4} - 2 \dfrac{a^{2/3} b^{2/3}}{6} + \dfrac{b^{4/3}}{9}$
$\qquad\qquad = \dfrac{1}{4} a^{4/3} - \dfrac{1}{3} a^{2/3} b^{2/3} + \dfrac{1}{9} b^{4/3}.$

20. $x^{-1} + x^{-1/2} y^{-1/2} + y^{-1}$
$\underline{x^{-1} + x^{-1/2} y^{-1/2} + y^{-1}}$
$x^{-2} + x^{-3/2} y^{-1/2} + x^{-1} y^{-1}$
$\quad - x^{-3/2} y^{-1/2} - x^{-1} y^{-1} - x^{-1/2} y^{-3/2}$
$\qquad\qquad + x^{-1} y^{-1} + x^{-1/2} y^{-3/2} + y^{-2}$
$\overline{x^{-2} \qquad\quad + x^{-1} y^{-1} \qquad\qquad + y^{-2}.}$

21. $(x^{-\frac{1}{2}} - y^{-\frac{1}{2}})^2 = x^{-1} - 2x^{-\frac{1}{2}}y^{-\frac{1}{2}} + y^{-1}$.

22. $\quad x^{\frac{2}{3}} - 4x^{\frac{1}{3}} + 2$
$\quad\;\; x \;\; - x^{\frac{1}{2}}$
$\overline{x^{\frac{5}{3}} - 4x^{\frac{4}{3}} + 2x}$
$\qquad\quad + 4x - x^2 - 2x^{\frac{1}{2}}$
$\overline{x^{\frac{5}{3}} - 4x^{\frac{4}{3}} + 6x - x^2 - 2x^{\frac{1}{2}}}.$

23. $\frac{1}{5}x^{\frac{9}{10}} - \frac{3}{10}x^{\frac{7}{10}} + \frac{1}{2}x^{\frac{3}{4}} - \frac{1}{3}x^{\frac{7}{12}} - \frac{3}{20}x^{\frac{9}{20}} + \frac{1}{4}x^{\frac{1}{2}}\;\Big|\;x^{\frac{1}{2}} - \frac{3}{5}x^{\frac{1}{5}}$
$\quad\quad -\frac{3}{10}x^{\frac{7}{10}} + \frac{1}{2}x^{\frac{3}{4}}\qquad\qquad\qquad\qquad\;\;\overline{\frac{1}{5}x^{\frac{2}{5}} - \frac{1}{3}x^{\frac{1}{3}} + \frac{1}{4}x^{\frac{1}{4}}}$
$\overline{\frac{1}{5}x^{\frac{9}{10}}\qquad\qquad\quad -\frac{1}{3}x^{\frac{7}{12}} - \frac{3}{20}x^{\frac{9}{20}} + \frac{1}{4}x^{\frac{1}{2}}}$
$\;\;\frac{1}{5}x^{\frac{9}{10}}\qquad\qquad\quad -\frac{1}{3}x^{\frac{7}{12}}$
$\qquad\qquad\qquad\qquad\qquad\qquad -\frac{3}{20}x^{\frac{9}{20}} + \frac{1}{4}x^{\frac{1}{2}}$
$\qquad\qquad\qquad\qquad\qquad\qquad -\frac{3}{20}x^{\frac{9}{20}} + \frac{1}{4}x^{\frac{1}{2}}.$
$\qquad\qquad\qquad\qquad\qquad\qquad\qquad\qquad 0$

24. $\quad x^2 + xy + y^2$
$\quad x + x^{\frac{1}{2}}y^{\frac{1}{2}} + y$
$\overline{x^3 + x^2y + xy^2}$
$\qquad\qquad\qquad + x^{\frac{5}{2}}y^{\frac{1}{2}} + x^{\frac{3}{2}}y^{\frac{3}{2}} + x^{\frac{1}{2}}y^{\frac{5}{2}}$
$\quad + x^2y + xy^2 \qquad\qquad\qquad\qquad\qquad\qquad + y^3$
$\overline{x^3 + 2x^2y + 2xy^2 + x^{\frac{5}{2}}y^{\frac{1}{2}} + x^{\frac{3}{2}}y^{\frac{3}{2}} + x^{\frac{1}{2}}y^{\frac{5}{2}} + y^3}.$

25. $\quad x^{\frac{2}{3}} + 2x^{\frac{1}{2}} + 3x^{\frac{1}{3}} + 2x^{\frac{1}{6}} + 1$
$\quad x^{\frac{1}{3}} - 2x^{\frac{1}{6}} + 1$
$\overline{x\;\; + 2x^{\frac{5}{6}} + 3x^{\frac{2}{3}} + 2x^{\frac{1}{2}} + \;\;x^{\frac{1}{3}}}$
$\qquad - 2x^{\frac{5}{6}} - 4x^{\frac{2}{3}} - 6x^{\frac{1}{2}} - 4x^{\frac{1}{3}} - 2x^{\frac{1}{6}}$
$\qquad\qquad + x^{\frac{2}{3}} + 2x^{\frac{1}{2}} + 3x^{\frac{1}{3}} + 2x^{\frac{1}{6}} + 1$
$\overline{x \qquad\qquad\quad - 2x^{\frac{1}{2}} \qquad\qquad\qquad + 1.}$

26.
$$x^2 + x^{\frac{3}{2}} - 1\tfrac{1}{3}x - \tfrac{2}{3}x^{\frac{1}{2}} + \tfrac{4}{9} \;\big|\; x + \tfrac{1}{2}x^{\frac{1}{2}} - \tfrac{2}{3}$$
$$\underline{x^2}$$
$$\underline{2x + \tfrac{1}{2}x^{\frac{1}{2}}\;\big|\; x^{\frac{3}{2}}}$$
$$\underline{x^{\frac{3}{2}} + \tfrac{1}{4}x}$$
$$2x + x^{\frac{1}{2}} - \tfrac{2}{3}\;\big|\; -1\tfrac{6}{12}x - \tfrac{2}{3}x^{\frac{1}{2}} + \tfrac{4}{9}$$
$$\underline{-1\tfrac{6}{12}x - \tfrac{2}{3}x^{\frac{1}{2}} + \tfrac{4}{9}.}$$

## ARITHMETICAL PROGRESSION.

2. $a = 2$, $d = 7$, $n = 100$; $\therefore l = a + (n-1)d = 695$.

3. $a = 1$, $d = 2$, $n = 100$; $\therefore l = a + (n-1)d = 199$.

Hence, $\quad S = \tfrac{1}{2}n(a + l) = 10000$.

4. $l = 70$, $d = 3$, $n = 21$,

$a = l - (n-1)d = 10$; $S = \tfrac{1}{2}n[2l - (n-1)d] = 840$.

5. $a = 10$, $d = -\tfrac{1}{3}$, $n = 21$ $\therefore l = a + (n-1)d = \tfrac{10}{3}$;

whence, $\quad S = \tfrac{1}{2}n[2l - (n-1)d] = 140$.

6. $d = 6$, $l = 185$, $S = 2945$

$$n = \frac{2l + d \pm \sqrt{(2l+d)^2 - 8dS}}{2d} = \frac{376 \pm 4}{12}$$

taking the lower sign, $n = 31$, $a = l - (n-1)d = 5$.

7. $a = 2$, $l = 5$, $n = 11$; $\therefore d = \dfrac{l-a}{n-1} = \dfrac{3}{10} = 0.3$.

8. $a = 1$, $d = 1$, $n = n$ $l = n$;

$\therefore S = \tfrac{1}{2}n[2a + (n-1)d] = \dfrac{(1+n)n}{2}$.

9. $a = 1$, $d = 2$, $n = n$; $\therefore S = \tfrac{1}{2}n[2a + (n-1)d] = n^2$

10. $a = 4$, $d = 4$, $n = 100$;

$\therefore S = \tfrac{1}{2}n[2a + (n-1)d] = 20200$

$20200 yds. = 11 mi.\ 840 yds.$

## GEOMETRICAL PROGRESSION.

**3.** $a = 2, \quad r = 3, \quad l = 39366,$

$$S = \frac{lr - a}{r - 1} = 59048.$$

**4.** $a = 1, \quad r = 2, \quad n = 12;$

$$\therefore \quad S = \frac{ar^n - a}{r - 1} = 4095; \quad l = ar^{n-1} = 2048.$$

**5.** $a = 1, \quad r = 2, \quad n = 12;$

$$\therefore \quad S = \frac{ar^n - a}{r - 1} = 4095; \quad 4095s. = £204 \; 15s.$$

**6.** $a = 1, \quad r = 3, \quad n = 10;$

$$\therefore \quad S = \frac{ar^n - a}{r - 1} = \frac{59048}{2} = 295.24.$$

$$l = ar^{n-1} = 196.83.$$

### INSERTING GEOMETRICAL MEANS.

**2.** $a = 2, \quad b = 486, \quad m = 4; \quad \therefore \quad r = \sqrt[5]{243} = 3:$
hence, the progression, $\quad 2 : 6 : 18 : 54 : 162 : 486.$

### SUMMATION OF SERIES.

**4.** $\quad p = 4, \quad q = 4, \quad n = 1, \; 5, \; 9, \; 13, \; \&c.$

1st auxiliary series, $\quad \dfrac{4}{1} + \dfrac{4}{5} + \dfrac{4}{9} + \dfrac{4}{13} + \dfrac{4}{17} + \cdots$

2d " " $\quad + \dfrac{4}{5} + \dfrac{4}{9} + \dfrac{4}{13} + \dfrac{4}{17} + \cdots + \dfrac{4}{4n - 3};$

$$S = \frac{1}{4}\left(4 - \frac{4}{4n - 3}\right); \text{ if } n = \infty, \quad S = 1.$$

## PILING BALLS.

1. $n = 15$; $\therefore S = \dfrac{n(n+1)(n+2)}{1\cdot 2\cdot 3} = 680.$

2. $n = 14$; $\therefore S = \dfrac{n(n+1)(2n+1)}{1\cdot 2\cdot 3} = 1015$;

   $n' = 5$, $S' = 55$; $\therefore S - S' = 960.$

3. $m = 30$, $n = 30$; $\therefore S = \dfrac{n(n+1)(1+2n+3n)}{1\cdot 2\cdot 3} = 23405.$

4. $m = 26$, $n = 20$; $\therefore S = 8330$; $m = 26$, $n' = 8$;

   $S' = 1140$; $\therefore S - S' = 7190.$

5. $n = 20$; $\therefore S = 1540$; $n' = 9$; $\therefore S' = 65$;

   $\therefore S - S' = 1475.$

6. $n = 15$; $\therefore S = 1240$; $n' = 5$; $\therefore S' = 55$;

   $\therefore S - S' = 1185.$

7. $m = 52$, $n = 40$; $\therefore S = 64780$; $m = 52$, $n' = 18$;

   $\therefore S' = 11001$; $\therefore S - S' = 53679.$

## EXPONENTIAL EQUATIONS.

These equations may be solved as the preceding ones have been, but it will be better to make use of the table of logarithms (P. 291).

$$8^x = 32; \text{ hence we have,}$$

$$x \log 8 = \log 32; \text{ or, } x = \frac{\log 32}{\log 8} = \frac{5}{3}$$

2. $\qquad 3^x = 15;$

taking the logarithms of both members,

$$x \log 3 = \log 15; \therefore x = \frac{\log 15}{\log 3} = \frac{1.176091}{0.477121} = 2.46.$$

3. $$10^x = 3;$$
taking the logarithms of both members,
$$x \log 10 = \log 3, \text{ or } x = \log 3 = 0.477121.$$

4. $$5^x = \tfrac{2}{3};$$
taking the logarithms of both members,
$$x \log 5 = \log \tfrac{2}{3} = \log 2 - \log 3;$$
$$\therefore x = \frac{\log 2 - \log 3}{\log 5} = \frac{-.176091}{.698970} = -0.25.$$

## THEORY OF EQUATIONS.

2. Two roots of the equation,
$$x^4 - 12x^3 + 48x^2 - 68x + 15 = 0,$$
are 3 and 5: what does the equation become when freed of them?

$$x^4 - 12x^3 + 48x^2 - 68x + 15 \,|\, \underline{x - 3}$$
$$\phantom{x^4 - 12x^3 + 48x^2 - 68x + 15\,|\,} x^3 - 9x^2 + 21x - 5$$

$$x^3 - 9x^2 + 21x - 5 \,|\, \underline{x - 5}$$
$$\phantom{x^3 - 9x^2 + 21x - 5\,|\,} x^2 - 4x + 1 = 0. \quad Ans.$$

3. A root of the equation,
$$x^3 - 6x^2 + 11x - 6 = 0,$$
is 1: what is the reduced equation?

$$x^3 - 6x^2 + 11x - 6 \,|\, \underline{x - 1}$$
$$\phantom{x^3 - 6x^2 + 11x - 6\,|\,} x^2 - 5x + 6 = 0. \quad Ans$$

4. Two roots of the equation,
$$4x^4 - 14x^3 - 5x^2 + 31x + 6 = 0,$$
are 2 and 3: find the reduced equation.

$$4x^4 - 14x^3 - 5x^2 + 31x + 6 \,|\, x - 2$$
$$\overline{4x^3 - 6x^2 - 17x - 3,}$$

$$4x^3 - 6x^2 - 17x - 3 \,|\, x - 3$$
$$\overline{4x^2 + 6x + 1 = 0.} \quad Ans$$

## FORMATION OF EQUATIONS.

**2.** What is the equation whose roots are 1, 2, and $-3$?

$$(x - 1)(x - 2)(x + 3) = x^3 - 7x + 6 = 0. \quad Ans.$$

**3.** What is the equation whose roots are 3, $-4$, $2 + \sqrt{3}$, and $2 - \sqrt{3}$?

$$(x - 3)(x + 4)(x - 2 - \sqrt{3})(x - 2 + \sqrt{3})$$
$$= x^4 - 3x^3 - 15x^2 + 49x - 12 = 0. \quad Ans.$$

**4.** What is the equation whose roots are $3 + \sqrt{5}$, $3 - \sqrt{5}$, and $-6$?

$$(x - 3 - \sqrt{5})(x - 3 + \sqrt{5})(x + 6) = x^3 - 32x + 24 = 0. \quad Ans.$$

**5.** What is the equation whose roots are 1, $-2$, 3, $-4$, 5, and $-6$?

$$(x - 1)(x + 2)(x - 3)(x + 4)(x - 5)(x + 6)$$
$$= x^6 + 3x^5 - 41x^4 - 87x^3 + 400x^2 + 444x - 720 = 0. \quad Ans.$$

**6.** What is the equation whose roots are .... $2 + \sqrt{-1}$, $2 - \sqrt{-1}$, and $-3$?

$$(x - 2 - \sqrt{-1})(x - 2 + \sqrt{-1})(x + 3)$$
$$= x^3 - x^2 - 7x + 15 = 0. \quad Ans.$$

## TRANSFORMATION OF EQUATIONS.

**2.** Transform the equation,

$$x^2 + 11x + 28 = 0,$$

into one whose roots are three times as great.

Make $x = \frac{7}{3}$; then we have,

$$\frac{y^2}{9} + \frac{11y}{3} + 28 = 0; \text{ or, } y^2 + 33y + 252 = 0.$$

3. Transform the equation,
$$x^5 + 3x^4 - 4x^3 + x^2 - x + 4 = 0,$$
into one whose roots are equal to those of the given equation with their signs changed.

Making $x = -y$, and dividing by $-1$,
$$y^5 - 3y^4 - 4y^3 - y^2 - y - 4 = 0.$$

1. Transform the equation,
$$x^3 + 3x^2 + x + \frac{1}{3} = 0,$$
into one whose roots shall be the reciprocals of those of the given equation.

Making $x = \frac{1}{y}$, substituting, and reducing,
$$y^3 + 3y^2 + 9y + 3 = 0.$$

2. Transform the equation,
$$x^4 + x^3 + 3x + 2 = 0,$$
into one whose roots shall be the reciprocals of those of the given equation.

Making $x = \frac{1}{y}$, substituting, and reducing,
$$y^4 + \frac{3}{2}y^3 + \frac{1}{2}y + \frac{1}{2} = 0.$$

3. Transform the equation,
$$x^2 + \frac{7}{2}x - \frac{7}{2} = 0,$$
into one whose roots shall be the reciprocals of those of the given equation.

Making $x = \frac{1}{y}$, substituting, and reducing,
$$y^2 - y - \frac{2}{7} = 0.$$

## TRANSFORMATION OF EQUATIONS BY SYNTHETICAL DIVISION.

4. Find the equation whose roots shall be less by 3 than the roots of the equation

$$x^4 - 3x^3 - 15x^2 + 49x - 12 = 0.$$

$$
\begin{array}{l}
1 - 3 - 15 + 49 - 12 \,\|\, 3 \\
\phantom{1} + 3 + \phantom{0}0 - 45 + 12 \\
\hline
\phantom{1} + 0 - 15 + \phantom{0}4, + \phantom{0}0 \\
\phantom{1} + 3 + \phantom{0}9 - 18 \\
\hline
\phantom{1} + 3 - \phantom{0}6, - 14 \\
\phantom{1} + 3 + 18 \\
\hline
\phantom{1} + 6 + 12 \\
\phantom{1} + 3 \\
\hline
1, + 9 \quad \therefore \quad y^4 + 9y^3 + 12y^2 - 14y = 0
\end{array}
$$

5. Find the equation whose roots shall be less by 10 than the roots of the equation

$$x^4 + 2x^3 + 3x^2 + 4x - 12340 = 0.$$

$$
\begin{array}{l}
1 + \phantom{0}2 + \phantom{00}3 + \phantom{000}4 - 12340 \,\|\, 10 \\
\phantom{1} + 10 + 120 + 1230 + 12340 \\
\hline
\phantom{1} + 12 + 123 + 1234, + \phantom{00}0 \\
\phantom{1} + 10 + 220 + 3430 \\
\hline
\phantom{1} + 22 + 343, + 4664 \\
\phantom{1} + 10 + 320 \\
\hline
\phantom{1} + 32, + 663 \\
\phantom{1} + 10 \\
\hline
1, + 42 \quad \therefore \quad y^4 + 42y^3 + 663y^2 + 4664y = 0.
\end{array}
$$

6. Find the equation whose roots shall be less by 2 than the roots of the equation

$$x^5 + 2x^3 - 6x^2 - 10x = 0.$$

$$
\begin{array}{l}
1+0+\phantom{0}2-\phantom{0}6-10+\phantom{0}0 \,\underline{|\,2} \\
\phantom{1}+2+\phantom{0}4+12+12+\phantom{0}4 \\ \hline
\phantom{1}+2+\phantom{0}6+\phantom{0}6+\phantom{0}2,+4 \\
\phantom{1}+2+\phantom{0}8+28+68 \\ \hline
\phantom{1}+4+14+34,+70 \\
\phantom{1}+2+12+52 \\ \hline
\phantom{1}+6+26,+86 \\
\phantom{1}+2+16 \\ \hline
\phantom{1}+8,+42 \\
\phantom{1}+2 \\ \hline
1,+10;\ \therefore\ y^5+10y^4+42y^3+86y^2+70y+4=0.
\end{array}
$$

## DISAPPEARANCE OF SECOND TERM.

2. Transform the equation

$$x^5 - 10x^4 + 7x^2 + 4x - 9 = 0$$

into an equation in which the second term shall be wanting.

Here we make the roots of the resulting equation greater than those of the given equation by $-\dfrac{P}{5}$, or $-2$ (Bourdon, Art. 266); that is, we make them less than those of the given equation by $+2$. Hence,

### OPERATION.

$$
\begin{array}{l}
1-10+\phantom{0}7+\phantom{0}0+\phantom{0}4-\phantom{0}9\,\underline{|\,2} \\
\phantom{1}+\phantom{0}2-16-18-38-64
\end{array}
$$

## TRANSFORMATION OF EQUATIONS.

| | | |
|---|---|---|
| 1st quotient, | $1 - 8 - 9 - 18 - 32, -73$ | 1st rem. |
| | $+ 2 - 12 - 42 - 120$ | |
| 2d quotient, | $1 - 6 - 21 - 60, -152$ | 2d rem. |
| | $+ 2 - 8 - 58$ | |
| 3d quotient, | $1 - 4 - 29, -118$ | 3d rem. |
| | $+ 2 - 4$ | |
| 4th quotient, | $1 - 2, -33$ | 4th rem. |
| | $+ 2$ | |
| 5th quotient, | $1, + 0$ | 5th rem. |

Hence, the transformed equation is

$$y^5 - 33y^3 - 118y^2 - 152y - 73 = 0.$$

3. Transform the equation

$$x^3 - 6x^2 + 7x - 10 = 0$$

into one whose second term shall be wanting.

Here we make the roots of the resulting equation greater than those of the given equation by $-\dfrac{P}{3} = -2$; that is, we make them less than those of the given equation by $+2$.

### OPERATION.

| | | |
|---|---|---|
| | $1 - 6 + 7 - 10 \underline{\,|\,2}$ | |
| | $+ 2 - 8 - 2$ | |
| 1st quotient, | $1 - 4 - 1, -12$ | 1st rem. |
| | $+ 2 - 4$ | |
| 2d quotient, | $1 - 2, -5$ | 2d rem. |
| | $+ 2$ | |
| 3d quotient, | $1, \ 0$ | 3d rem. |

Hence, the transformed equation is
$$y^3 - 5y - 12 = 0.$$

4. Transform the equation,
$$x^3 + 9x^2 - x + 4 = 0$$
into one whose second term shall be wanting.

Here we make the roots of the resulting equation greater than those of the given equation by $+\dfrac{P}{3} = 3$; that is, we make them greater than those of the given equation by 3. Therefore, the synthetical divisor is $-3$.

OPERATION.

```
                 1 + 9 —  1 +  4 |— 3
                   — 3 — 18 + 57
1st quotient,    1 + 6 — 19, + 61    1st rem.
                 1 — 3 —  9
2d quotient,     1 + 3, — 28    2d rem.
                   — 3
3d quotient,     1, + 0    3d rem.
```

Hence, the transformed equation is
$$y^3 - 28y + 61 = 0.$$

5. Transform the equation
$$x^4 - 8x^3 + 7x^2 + 3x + 4 = 0$$
into one whose second term shall be wanting.

Here we make the roots of the resulting equation greater than those of the given equation by $-\dfrac{P}{4} = -2$; that is, we make them less than those of the given equation by $+2$; hence, the synthetical divisor is $+2$.

OPERATION.

$$1 - 8 + 7 + 3 + 4 \underline{|2}$$
$$+ 2 - 12 - 10 - 14$$

1st quotient, $\quad 1 - 6 - 5 - 7, -10 \quad$ 1st rem.
$$\phantom{1 - 6 - 5} + 2 - 8 - 26$$

2d quotient, $\quad 1 - 4 - 13, -33 \quad$ 2d rem.
$$\phantom{1 - 4 - 13} + 2 - 4$$

3d quotient, $\quad 1 - 2, -17 \quad$ 3d rem.
$$\phantom{1 - 2} + 2$$

4th quotient, $\quad 1, + 0 \quad$ 4th rem.

$$\therefore\ y^4 - 17y^2 - 33y - 10 = 0.$$

## EQUAL ROOTS.

4. What are the equal factors of the equation

$$x^7 - 7x^6 + 10x^5 + 22x^4 - 43x^3 - 35x^2 + 48x + 36 = 0?$$

The first derived polynomial is

$$7x^6 - 42x^5 + 50x^4 + 88x^3 - 129x^2 - 70x + 48,$$

and the common divisor between it and the first member of the given equation, is

$$x^4 - 3x^3 - 3x^2 + 7x + 6.$$

The equation

$$x^4 - 3x^3 + 3x^2 + 7x + 6 = 0,$$

cannot be solved directly, but by applying to it the method of equal roots; that is, by seeking for a common divisor between its first member and its derived polynomial,

$$4x^3 - 9x^2 - 6x + 7,$$

we find such divisor to be $x + 1$; hence, $x + 1$ is *twice* a factor of the first derived polynomial, and *three times* a factor of the first member of the given equation (Art. 271).

Dividing,

$$x^4 - 3x^3 - 3x^2 + 7x + 6 = 0, \quad \text{by} \quad (x+1)^2 = x^2 + 2x + 1,$$

we have, $\quad x^2 - 5x + 6,$

which being placed equal to 0, gives the two roots

$$x = 2 \quad \text{and} \quad x = 3,$$

and the two factors, $x - 2 \quad$ and $\quad x - 3.$

Therefore, $(x - 2)$ and $(x - 3)$, each enters twice as a factor of the given equation; hence, the factors are

$$(x - 2)^2 (x - 3)^2 (x + 1)^3. \quad \textit{Ans.}$$

5. What are the equal factors of the equation

$$x^7 - 3x^6 + 9x^5 - 19x^4 + 27x^3 - 33x^2 + 27x - 9 = 0?$$

The first derived polynomial is

$$7x^6 - 18x^5 + 45x^4 - 76x^3 + 81x^2 - 66x + 27,$$

and the common divisor between it and the first member of the given equation, is

$$x^4 - 2x^3 + 4x^2 - 6x + 3.$$

The equation

$$x^4 - 2x^3 + 4x^2 - 6x + 3,$$

cannot be solved directly, but by applying to it the method of equal roots, as in the last example, we find the derived polynomial to be

$$4x^3 - 6x^2 + 8x - 6$$

and the common divisor to be $x - 1$; hence, $(x - 1)$ enters twice as a factor into the derived polynomial, and *three* times as a factor into the first member of the given equation.

Dividing

$$x^4 - 2x + 4x^2 - 6x + 3 \quad \text{by} \quad (x - 1)^2 = x^2 - 2x + 1,$$

we have for a quotient $x^2 + 3$;

hence, $(x^2 + 3)$ enters twice as a factor into the first member of the given equation; hence the factors are

$$(x — 1)^3 (x + 3)^2.$$

## SUPERIOR LIMIT IN ENTIRE ROOTS.

2 What is the superior limit of the positive roots in the equation

$$x^5 — 3x^4 — 8x^3 — 25x^2 + 4x — 39 = 0 ?$$

Recollecting that if we use $x$ for $x'$ (Art. 285), we have

$X = x^5 — 3x^4 — 8x^3 — 25x^2 + 4x — 39$

$Y = 5x^4 — 12x^3 — 24x^2 — 50x + 4,$

$Z = 20x^3 — 36x^2 — 48x — 50,$

$V = 60x^2 — 72x — 48,$

$W = 120x — 72.$

$T = 120$

The least whole number that will render all these derived polynomials positive is 6; hence, 6 is the superior limit.

3. What is the superior limit of the positive roots in the equation
$$x^5 — 5x^4 — 13x^3 — 17x^2 — 69 = 0.$$

The process of solution is the same as in the last example, and the limit is found to be 7.

### COMMENSURABLE ROOTS.

2. What are the entire roots of the equation

$$x^4 — 5x^3 + 25x — 21 = 0 ?$$

The divisors of the last term are $+ 1, — 1, + 3, — 3, + 7, — 7$, $+ 21$, and $— 21: L = 22; — L'' = — 4.$

$+21,\ +7,\ +3,\ +1,\ -1,\ -3,\ -$
$-1,\ -3,\ -7,\ -21,\ +21,\ +7,\ +$
$+24,\ +22,\ +18,\ +4,\ +46,\ +32,\ +$
$+3,\ +6,\ +4,\ -46,\qquad\qquad -$
$+2,\ +4,\ +46,$
$-3,\ -1,\ +41,$
$-1,\ -1;\ -41,$

therefore, $+3$ and $+1$ are the two entire roots. Dividing the first member of the equation by the product of the factors

$$(x-3)(x-1) = x^2 - 4x + 3,$$

we have $\qquad x^2 - x - y = 0.$

NOTE.—In the 4th line we add the co-efficient of $x^3$, which is 0, and then divide by the divisors, and thus obtain the 6th line.

3. What are the entire roots of the equation

$$15x^5 - 19x^4 + 6x^3 + 15x^2 - 19x + 6 = 0\ ?$$

$+6,\ +3,\ +2,\ +1,\ -1,\ -2,\ -3,\ -6,$
$+1,\ +2,\ +3,\ +6,\ -6,\ -3,\ -2,\ -1,$
$-18,\ -17,\ -16,\ -13.\ -25,\ -22,\ -21,\ -20,$
$-3,\qquad\ \ -8,\ -13,\ +25,\ +11,\ +7,$
$+12,\qquad\ +7,\ +2,\ +40,\ +26,\ +22,$
$+2,\qquad\qquad\ \ +2,\ -40,\ -13,$
$+8,\qquad\qquad\ \ +8,\ -34,\ -7,$
$\qquad\qquad\qquad\ \ +8,\ +34,$
$\qquad\qquad\qquad\ -11,\ +15,$
$\qquad\qquad\qquad\ -11,\ -15,$
$\qquad\qquad\qquad\qquad\ \ +15:$

hence, there is but one entire root, which is $-1$

4. What are the entire roots of the equation

$$9x^6 + 30x^5 + 22x^4 + 10x^3 + 17x^2 - 20x + 4 = 0?$$

This is worked like the preceding example, giving the entire root, $-2$. Then dividing the equation by $x+2$, we find a new one, which has a root, $-2$.

### NUMBER OF REAL ROOTS.

3. What is the number of real roots of the equation

$$x^3 - 5x^2 + 8x - 1 = 0?$$

By finding the expressions which indicate, by their change of sign, the existence of real roots (Art. 293 and Example 1), we have

$$X = x^3 - 5x^2 + 8x - 1$$
$$X_1 = 3x^2 - 10x + 8$$
$$X_2 = 2x - 31$$
$$X_3 = -2295$$

$x = -\infty$ gives $- + - -$ 2 variations,
$x = +\infty$ gives $+ + + -$ 1 variation;

hence, there is one real and two imaginary roots (Art. 293).

For $x = 0$, we have $- + - -$ 2 variations,
for $x = 1$, " $+ + - -$ 1 variation;

hence, the real root lies between 0 and $+1$.

4. Find the number, places, and limits of the real roots of

$$x^4 - 8x^3 + 14x^2 + 1x - 8 = 0.$$

For solution, see Example 3, page 141 of Key.

5. Find the number, places, and limits of the real roots of the equation

$$x^3 - 23x - 24 = 0.$$

$$X = x^3 - 23x - 24$$
$$X_1 = 3x^2 - 23$$
$$X_2 = 23x + 36$$
$$X_3 = 8279.$$

For $x = -\infty$, we have, $- + - +$, 3 variations.
For $x = +\infty$, we have, $+ + + +$, no variations.
Hence, there are 3 real roots, which are easily placed.

6. In the sixth example, we have,

$$X = x^3 + \tfrac{1}{2}x^2 - 2x - 5$$
$$X_1 = 3x^2 + 3x - 2$$
$$X_2 = 11x + 28$$
$$X_3 = -1186.$$

Hence, there is but one real root, and that lies between the limits 1 and 2.

## CUBIC EQUATIONS.

1. What are the roots of the equation
$$x^3 - 6x^2 + 3x = 18 \quad \cdot \quad \cdot \quad \cdot \quad (1)?$$

Transforming so as to make the second term disappear,
$$x^3 - 9x - 28 = 0 \quad \cdot \quad \cdot \quad \cdot \quad (2)$$
$$p = -9 \quad q = -28;$$

substituting in Cardan's formula, and reducing,
$$x = 4.$$

But the roots of the given equation are greater than those of equation (2) by 2; hence $x = 6$.

Transposing 18 in equation (1), and dividing both numbers by $x - 6$, we find

$$x^2 + 3 = 0 \quad \therefore \quad x = \pm \sqrt{-3};$$

hence the three roots are $\quad 6, \sqrt{-3}, -\sqrt{-3}.$

2. What are the roots of the equation

$$x^3 - 9x^2 + 28x = 30 \quad \cdots \quad (1)?$$

Transforming, we find

$$x^3 + x = 0 \ (2) \quad \therefore \quad x = 0 \text{ and } x = \pm \sqrt{-1}.$$

But the roots of (1) are greater than those of (2) by 3;

hence the roots of (1) are, $\quad 3, \ 3 + \sqrt{-1}$ and $3 - \sqrt{-1}.$

3. $\qquad x^3 - 7x + 14 = 20 \quad \ldots \ldots \quad (1)$

Transforming (Bourdon, Art. 266), we have,

$$x^3 - \frac{7}{3}x - \frac{344}{7} = 0 \ (2); \quad \text{hence,} \quad p = -\frac{7}{3}, \quad q = -\frac{344}{27}.$$

Substituting in Cardan's formula, we have $\frac{8}{3}$, the real root. But the roots of (1) are greater by $\frac{7}{3}$ than those of (2); hence, in (1) $x = 5.$

Transposing and dividing by $x - 5,$ we have,

$$x^2 - 2x + 4 = 0;$$

hence, the required roots are, $5; \ 1 + \sqrt{-3},$ and $1 - \sqrt{-3}.$

## HORNER'S METHOD OF SOLVING NUMERICAL EQUATIONS.

1. $\qquad x^3 + x^2 + x - 100 = 0.$

By Sturm's Rule, we find

$$X = x^3 + x^2 + x - 100,$$
$$X_1 = 3x^2 + 2x + 1$$
$$X_2 = -4x + 899$$
$$X_3 = -2409336.$$

For $x = -\infty$, $- + + -$, 2 variations,
for $x = +\infty$, $+ + - -$, 1 variation;
hence, there is but one real root.

For $x = 4$, $- + + -$, 2 variations,
for $x = 5$, $+ + + -$, 1 variation;
hence, the real root lies between 4 and 5.

2. $\quad x^4 - 12x^2 + 12x - 3 = 0.$

By Sturm's Rule,
$$X = x^4 - 12x^2 + 12x - 3,$$
$$X_1 = 4x^3 - 24x + 12, \text{ or } x^3 - 6x + 3,$$
$$X_2 = 2x^2 - 3x + 1,$$
$$X_3 = 13x - 9,$$
$$X_4 = 20.$$

For $x = -\infty$, $+ - + - +$, 4 variations,
for $x = +\infty$, $+ + + + +$, 0 variation;
hence, there are 4 real roots.

For $x = -4$, $+ - + - +$, 4 variations,
for $x = -3$, $- - + - +$, 3 variations,
for $x = 0$, $- + + - +$, 3 variations,
for $x = +1$, $- - \mp + +$, 1 variation,
for $x = +2$, $- - + + +$, 1 variation,
for $x = +3$, $+ + + + +$, 0 variation;

hence, one of the roots lies between $-4$ and $-3$, two between 0 and 1, and the remaining root lies between 2 and 3.

3. $\qquad x^4 - 8x^3 + 14x^2 + 4x - 8 = 0.$

By Sturm's Rule,
$$X = x^4 - 8x^3 + 14x^2 + 4x - 8,$$
$$X_1 = 4x^3 - 24x^2 + 28x + 4, \text{ or } x^3 - 6x^2 + 7x + 1,$$
$$X_2 = 5x^2 - 17x + 6,$$
$$X_3 = 76x - 103,$$
$$X_4 = 45475.$$

| | | | |
|---|---|---|---|
| For | $x = -\infty$, | $+ - + - +$ | 4 variations, |
| for | $x = +\infty$, | $+ + + + +$ | 0 variation; |

hence, the equation has 4 real roots.

| | | | |
|---|---|---|---|
| For | $x = -1$ | $+ - + - +$ | 4 variations. |
| for | $x = 0$ | $- + + - +$ | 3 variations, |
| for | $x = +1$ | $+ + - - +$ | 2 variations, |
| for | $x = +2$ | $+ - - + +$ | 2 variations, |
| for | $x = +3$ | $- - \pm + +$ | 1 variation, |
| for | $x = 5$ | $- - + + +$ | 1 variation, |
| for | $x = 6$ | $+ + + + +$ | 0 variation; |

hence, one root is between $-1$ and 0, one between 0 and 1, one between 2 and 3, and one between 5 and 6.

4. $\qquad x^5 - 10x^3 + 6x + 1 = 0.$

By Sturm's Rule,
$$X = x^5 - 10x^3 + 6x + 1,$$
$$X_1 = 5x^4 - 30x^2 + 6,$$
$$X_2 = 20x^3 - 24x - 5,$$
$$X_3 = 96x^2 - 5x - 24,$$
$$X_4 = 43651x + 10920,$$
$$X_5 = 32335636224.$$

For $x = -\infty$    $- + - + - +$    5 variations,
for $x = +\infty$    $+ + + + + +$    0 variation;

hence, the equation has 5 real roots.

For $x = -4$    $- + - + - +$    5 variations,
for $x = -3$    $+ + - + - +$    4 variations,
for $x = -1$    $+ - - + - +$    4 variations,
for $x = 0$    $+ + - - + +$    2 variations,
for $x = +1$    $- - - + + +$    1 variation,
for $x = +3$    $- - + + + +$    1 variation;
for $x = +4$    $+ + + + + +$    0 variation;

hence,    one root lies between $-4$ and $-3$,
two roots lie between $-1$ and $0$,
one root lies between $0$ and $+1$,
one root lies between $3$ and $4$.

# APPENDIX.

GENERAL SOLUTION OF TWO SIMULTANEOUS EQUATIONS OF THE FIRST DEGREE.

**1.** Take the equations,

$$ax + by = c \quad \ldots \quad (1),$$
$$a'x + b'y = c' \quad \ldots \quad (2);$$

multiply both members of (1) by $b'$ and of (2) by $b$, then subtracting and factoring, we find

$$(ab' - a'b)x = b'c - bc';$$
$$\therefore \quad x = \frac{b'c - bc'}{ab' - a'b} \quad \ldots \quad (3).$$

In like manner,
$$y = \frac{ac' - a'c}{ab' - a'b} \quad \ldots \quad (4).$$

By means of formulas (3) and (4) any two simultaneous equations of the forms (1) and (2) may be solved.

Thus,
$$4x + 3y = 31,$$
$$3x + 2y = 22:$$

by comparison with (1) and (2),

$$a = 4, \quad b = 3, \quad c = 31, \quad a' = 3, \quad b' = 2, \quad c' = 22;$$

by substitution in (3) and (4),

$$x = \frac{62 - 66}{8 - 9} = 4, \quad y = \frac{88 - 93}{8 - 9} = 5.$$

## APPENDIX.

### EXAMPLES.

1. Given $\left\{ \begin{array}{l} \frac{x}{3} + \frac{y}{4} = 2 \\ 3x + 4y = 25 \end{array} \right\}$ to find $x$ and $y$.

By comparison with (1) and (2),

$$a = \tfrac{1}{3}, \quad b = \tfrac{1}{4}, \quad c = 2,$$
$$a' = 3, \quad b' = 4, \quad c' = 25;$$

by substitution in (3) and (4),

$$x = \frac{8 - 6\tfrac{1}{4}}{1\tfrac{1}{3} - \tfrac{3}{4}} = 3, \quad y = \frac{8\tfrac{1}{3} - 6}{1\tfrac{1}{3} - \tfrac{3}{4}} = 4.$$

2. Given $\left\{ \begin{array}{l} 11x - 5y = -1 \\ -5x + 16y = 124 \end{array} \right\}$ to find $x$ and $y$:

by comparison,

$$a = 11, \quad b = -5, \quad c = -1,$$
$$a' = -5, \quad b' = 16, \quad c' = 124;$$

by substitution,

$$x = \frac{-16 + 620}{176 - 25} = 4, \quad y = \frac{1364 - 5}{176 - 25} = \ .$$

### GENERAL SOLUTION OF THREE SIMULTANEOUS EQUATIONS OF THE FIRST DEGREE.

2. Take the equations,

$$ax + by + cz = d \quad \cdots \quad (1),$$
$$a'x + b'y + c'z = d' \quad \cdots \quad (2),$$
$$a''x + b''y + c''z = d'' \quad \cdots \quad (3).$$

From (1) and (2) we obtain, by eliminating $z$,

$$(c'a - ca')x + (c'b - cb')y = c'd - cd' \quad \cdots \quad (4).$$

In like manner, from (1) and (3),

$$(c''a - ca'')x + (c''b - cb'')y = c''d - cd'' \quad \cdot \cdot \cdot \quad (5);$$

combining (4) and (5) and eliminating $y$, we find

$$x = \frac{(c''b - cb'')(c'd - cd') - (c'b - cb')(c''d - cd'')}{(c'a - ca')(c''b - cb'') - (c''a - ca'')(c'b - cb')} \quad \cdot \cdot \quad (6).$$

In like manner,

$$y = \frac{(c'a - ca')(c''d - cd'') - (c''a - ca'')(c'd - cd')}{(c'a - ca')(c''b - cb'') - (c''a - ca'')(c'b - cb')} \quad \cdot \cdot \quad (7),$$

$$z = \frac{(a''b - ab'')(a'd - ad') - (a'b - ab')(a''d - ad'')}{(c'a - ca')(b''a - ba'') - (c''a - a''c)(b'a - ba')} \quad \cdot \quad (8)$$

Formulas (6), (7) and (8) enable us to solve all groups of simultaneous equations of the form of (1), (2) and (3). Thus,

$$2x + 3y + 4z = 29,$$
$$3x + 2y + 5z = 32,$$
$$4x + 3y + 2z = 25:$$

by comparison with (1), (2) and (3),

$$a = 2, \quad b = 3, \quad c = 4, \quad d = 29,$$
$$a' = 3, \quad b' = 2, \quad c' = 5, \quad d' = 32,$$
$$a'' = 4, \quad b'' = 3, \quad c'' = 2, \quad d'' = 25:$$

by substitution in (6), (7) and (8),

$$x = \frac{(6 - 12)(145 - 128) - (15 - 8)(58 - 100)}{(10 - 12)(6 - 12) - (4 - 16)(15 - 8)}$$

$$= \frac{-102 + 294}{12 + 84} = \frac{192}{96} = 2,$$

$$y = \frac{(10-12)(58-100) - (4-16)(145-128)}{(10-12)(6-12) - (4-6)(15-8)}$$

$$= \frac{84 + 204}{12 + 84} = \frac{288}{96} = 3,$$

$$z = \frac{(12-6)(87-64) - (9-4)(116-50)}{(10-12)(6-12) - (4-16)(4-9)}$$

$$= \frac{138 - 330}{+12 - 60} = \frac{192}{48} = 4.$$

### EXAMPLES.

1. Given $\begin{cases} x + y + z = 90 \\ 2x - 3y = -20 \\ 2x - 4z = -30 \end{cases}$ to find $x$, $y$ and $z$.

By comparison with (1), (2) and (3),

$a = 1,\quad b = 1,\quad c = 1,\quad d = 90$

$a' = 2,\quad b' = -3,\quad c' = 0,\quad d' = -20,$

$a'' = 2,\quad b'' = 0,\quad c'' = -4,\quad d'' = -30;$

by substitution,

$$x = \frac{-4 \times 20 - 3 \times -330}{-2 \times -4 + 6 \times 3} = \frac{910}{26} = 35,$$

$$y = \frac{-2 \times -330 + 6 \times 20}{-2 \times 4 + 6 \times 3} = \frac{780}{26} = 30,$$

$$z = \frac{2 \times 200 - 5 \times 210}{-2 \times -2 + 6 \times -5} = \frac{-650}{-26} = 25.$$

2. Given $\begin{cases} x + y + z = 6 \\ x + 2y + 3z = 14 \\ 3x - y + 4z = 13 \end{cases}$ to find $x$, $y$ and $z$:

by comparison,

$$a = 1, \quad b = 1, \quad c = 1, \quad d = 6,$$
$$a' = 1, \quad b' = 2, \quad c' = 3, \quad d' = 14,$$
$$a'' = 3, \quad b'' = -1, \quad c'' = 4, \quad d'' = 13:$$

by substitution,

$$x = \frac{5 \times 4 - 11}{2 \times 5 - 1} = 1; \quad y = \frac{2 \times 11 - 4}{2 \times 5 - 1} = 2; \quad z = \frac{4 \times -8 + 1 \times 5}{2 \times -4 - 1} = 3$$

ELIMINATION BY THE METHOD OF ARBITRARY MULTIPLIERS.

**3.** There is a method of elimination by means of arbitrary quantities that will often be found useful, particularly in the higher investigations of applied mathematics. It consists in multiplying both members of one of the given equations by an arbitrary quantity, then adding the resulting equation to the second of the given equations, member to member, after which such a value is to be assigned to the arbitrary quantity as will reduce the co-efficient of the quantity to be eliminated to 0.

To illustrate, let us take the two simultaneous equations,

$$ax + by = c \quad \cdot \quad \cdot \quad \cdot \quad (1),$$
$$a'x + b'y = c' \quad \cdot \quad \cdot \quad \cdot \quad (2);$$

multiplying both members of (1) by $n$, which is entirely arbitrary, we have

$$nax + nby = nc \quad \cdot \quad \cdot \quad \cdot \quad (3);$$

adding (2) and (3), member to member, and factoring,

$$(na + a')x + (nb + b')y = nc + c' \quad \cdot \quad \cdot \quad \cdot \quad (4).$$

If it be required to eliminate $y$, place

$$nb + b' = 0; \quad \therefore \quad n = -\frac{b'}{b};$$

substituting this in (4) and reducing, we find

$$x = \frac{-\frac{b'}{b}c + c'}{-\frac{b'}{b}a + a'} = \frac{b'c - bc'}{ab' - a'b} \quad \cdots \quad (5).$$

If it be required to eliminate $x$, place

$$na + a' = 0; \quad \therefore \quad n = -\frac{a'}{a};$$

substituting in (4) and reducing, we find

$$y = \frac{-\frac{a'}{a}c + c'}{-\frac{a'}{a}b + b'} = \frac{ac' - a'c}{ab' - a'b} \quad \cdots \quad (6).$$

These values of $x$ and $y$ correspond to those already deduced by previous methods.

As an example, let it be required to find the values of $x$ and $y$ from the equations

$$3x - y = 5 \quad \cdots \quad (1),$$
$$7x + 3y = 33 \quad \cdots \quad (2);$$

multiplying both members of (1) by $n$,

$$3nx - ny = 5n \quad \cdots \quad (3);$$

adding (2) and (3), member to member,

$$(3n + 7)x + (3 - n)y = 5n + 33 \quad \cdots \quad (4):$$

1st. Assume $\quad 3 - n = 0; \quad \therefore \quad n = 3;$

substituting in (4), and reducing,

$$x = \frac{15 + 33}{9 + 7} = 3.$$

2d. Assume $\quad 3n + 7 = 0; \quad \therefore \quad n = -\frac{7}{3};$

substituting in (4), and reducing,

$$y = \frac{-\frac{35}{3} + 33}{3 + \frac{7}{3}} = 4.$$

### EXAMPLES.

1. Given $\left\{\begin{array}{l} x + \frac{1}{2}y = 14 \\ \frac{1}{2}x - \frac{1}{6}y = 2 \end{array}\right\}$ to find $x$ and $y$.

Multiplying both members of the first equation by $n$ and adding to the second, member by member,

$$\left(n + \frac{1}{2}\right)x + \left(\frac{n}{2} - \frac{1}{6}\right)y = 14n + 2;$$

making $\quad n = -\frac{1}{2} \quad$ and reducing,

$$y = 12;$$

making $\quad n = \frac{1}{3} \quad$ and reducing,

$$x = 8.$$

2. Given $\left\{\begin{array}{l} x - \frac{y-2}{7} = 5 \\ 4y - \frac{x+10}{3} = 3 \end{array}\right\}$ to find $x$ and $y$.

Reducing and transposing,

$$7x - y = 33 \quad \cdots \quad (1),$$
$$12y - x = 19 \quad \cdots \quad (2):$$

multiplying by $n$, and adding and factoring,

9

$$(7n - 1) x - (n - 12) y = 33n + 19:$$

making $\quad n = \dfrac{1}{7}, \quad$ we find $\quad y = 2;$

making $\quad n = 12, \quad$ we find $\quad x = 5.$

3. Given $\quad \left\{ \begin{array}{l} \dfrac{9}{x} + \dfrac{6}{y} = 36 \\ \dfrac{14}{x} - \dfrac{6}{y} = 10 \end{array} \right\} \quad$ to find $x$ and $y.$

Multiplying by $n$, adding and factoring,

$$(9n + 14)\dfrac{1}{x} + (6n - 6)\dfrac{1}{y} = 36n + 10;$$

making $\quad n = -\dfrac{14}{9}, \quad$ we have $\quad \dfrac{1}{y} = 3; \quad \therefore y = \dfrac{1}{3};$

making $\quad n = 1, \quad$ we have $\quad \dfrac{1}{x} = 2; \quad \therefore x = \dfrac{1}{2}.$

## MISCELLANEOUS GROUPS OF SIMULTANEOUS EQUATIONS OF THE FIRST DEGREE.

1. Given $\quad \left\{ \begin{array}{l} \dfrac{1}{3x} + \dfrac{1}{5y} = \dfrac{2}{9} \\ \dfrac{1}{5x} + \dfrac{1}{3y} = \dfrac{1}{4} \end{array} \right\} \quad$ to find $x$ and $y.$

Combining and eliminating $\dfrac{1}{x}$,

$$\left(\dfrac{1}{25} - \dfrac{1}{9}\right)\dfrac{1}{y} = \dfrac{2}{45} - \dfrac{1}{12};$$

whence, $\quad \dfrac{1}{y} = \dfrac{35}{64}; \quad \therefore y = 1\dfrac{29}{35}.$

Combining and eliminating $\dfrac{1}{y}$,

$$\left(\dfrac{1}{9}-\dfrac{1}{25}\right)\dfrac{1}{x}=\dfrac{2}{27}-\dfrac{1}{12};$$

whence, $\qquad \dfrac{1}{x}=\dfrac{65}{192}; \quad \therefore\ x=2\dfrac{62}{65}.$

2. Given $\qquad \begin{cases} \dfrac{x}{3}+2y=5 \\ \dfrac{2x-1}{5}-y+1=0 \end{cases}$ to find $x$ and $y$;

Clearing of fractions and reducing,

$$x+6y=15 \quad \cdots \quad (1),$$
$$2x-5y=-4 \quad \cdots \quad (2):$$

combining and eliminating $x$,

$$17y=34; \quad \therefore\ y=2;$$

substituting in (1), $\qquad x=3.$

3. Given $\qquad \begin{cases} x-\dfrac{y-2}{7}=5 \\ 4y-\dfrac{x+10}{3}=3 \end{cases}$ to find $x$ and $y$.

Clearing of fractions and reducing,

$$7x-y=33 \quad \cdots \quad (1),$$
$$12y-x=19 \quad \cdots \quad (2);$$

combining and eliminating $x$,

$$83y=166; \quad \therefore\ y=2;$$

substituting in (1), $\qquad x=5.$

**4.** Given $\left\{\begin{array}{l}\dfrac{5x-4}{6}+2y=24\\ \dfrac{20-2y}{5}+5x=40\frac{2}{5}\end{array}\right\}$ to find $x$ and $y$.

Clearing of fractions and reducing,

$$5x + 12y = 148,$$
$$25x - 2y = 182;$$

combining and eliminating $y$,

$$155x = 1240; \quad \therefore \; x = 8;$$

substituting, we find $\quad y = 9.$

**5.** Given $\left\{\begin{array}{l}\dfrac{x}{2}+\dfrac{y}{3}=12-\dfrac{z}{6}\\ \dfrac{y}{2}+\dfrac{z}{3}=8+\dfrac{x}{6}\\ \dfrac{x}{2}+\dfrac{z}{3}=10\end{array}\right\}$ to find $x$, $y$ and $z$.

Clearing of fractions and reducing,

$$3x + 2y + z = 72 \cdots (1),$$
$$-x + 3y + 2z = 48 \cdots (2),$$
$$3x + \phantom{3y +} 2z = 60 \cdots (3);$$

combining (1) and (2), eliminating $y$,

$$11x - z = 120 \cdots (4);$$

combining (3) and (4), eliminating $z$,

$$25x = 300; \quad \therefore \; x = 12;$$

by substitution in (3), $\quad z = 12,$

" " (1), $\quad y = 12.$

6. Given
$$\begin{cases} \frac{x}{2} + \frac{y}{3} + \frac{z}{7} = 22 \\ \frac{x}{3} + \frac{y}{5} + \frac{z}{2} = 31 \\ \frac{x}{6} + \frac{y}{3} + \frac{z}{6} = 19 \end{cases}$$ to find $x$, $y$ and $z$

Clearing of fractions,

$$21x + 14y + 6z = 924 \quad \cdots \quad (1),$$
$$10x + 6y + 15z = 930 \quad \cdots \quad (2),$$
$$x + 2y + z = 114 \quad \cdots \quad (3);$$

combining (1) and (3), eliminating $z$,

$$15x + 2y = 240 \quad \cdots \quad (4);$$

combining (2) and (3), eliminating $z$,

$$5x + 24y = 780 \quad \cdots \quad (5);$$

combining (4) and (5), eliminating $x$,

$$70y = 2100; \quad \therefore \quad y = 30;$$

from (5), $x = 12$; from (3), $z = 42$.

7. Given
$$\begin{cases} 2x - 3y + 2z = 13 \quad \cdots \quad (1) \\ 2v - x = 15 \quad \cdots \quad (2) \\ 2y + z = 7 \quad \cdots \quad (3) \\ 5y + 3v = 32 \quad \cdots \quad (4) \end{cases}$$ to find $x$, $y$, $z$ and $v$.

Combining (2) and (4), eliminating $v$,

$$10y + 3x = 19 \quad \cdots \quad (5);$$

combining (5) and (1), eliminating $x$,

$$29y - 6z = -1 \quad \cdots \quad (6);$$

combining (6) and (3), eliminating $z$,
$$41y = 41; \quad \therefore \quad y = 1:$$
by successive substitutions, $\quad z = 5, \quad x = 3, \quad v = 9.$

8. Given
$$\begin{cases} \dfrac{2}{x} + \dfrac{3}{y} = \dfrac{1}{12} + \dfrac{4}{z} \\ \dfrac{3}{x} + \dfrac{5}{z} = \dfrac{19}{24} + \dfrac{4}{y} \\ \dfrac{7}{y} + \dfrac{5}{z} = \dfrac{3}{8} + \dfrac{5}{x} \end{cases}$$
to find $x$, $y$ and $z$.

Transposing, reducing, &c.,
$$2\frac{1}{x} + 3\frac{1}{y} - 4\frac{1}{z} = \frac{2}{24} \quad \cdots \quad (1),$$
$$3\frac{1}{x} - 4\frac{1}{y} + 5\frac{1}{z} = \frac{19}{24} \quad \cdots \quad (2),$$
$$-5\frac{1}{x} + 7\frac{1}{y} + 5\frac{1}{z} = \frac{9}{24} \quad \cdots \quad (3):$$

combining (1) and (3), also (2) and (3), eliminating $\dfrac{1}{z}$,
$$-10\frac{1}{x} + 43\frac{1}{y} = \frac{46}{24} \quad \cdots \quad (4),$$
$$8\frac{1}{x} - 11\frac{1}{y} = \frac{10}{24} \quad \cdots \quad (5);$$

combining and eliminating $\dfrac{1}{x}$,
$$234\frac{1}{y} = \frac{468}{24}; \quad \therefore \quad \frac{1}{y} = \frac{2}{24}, \quad \text{or} \quad y = 12;$$

substituting in (5), $\quad x = 6;$ whence, $z = 8.$

9. Given
$$\begin{cases} \dfrac{10 + 6y - 4x}{6x - 9y + 3} = \dfrac{4}{3} \\ \dfrac{126 + 8x - 17y}{100 - 12x + 7y} = \dfrac{35}{13} \end{cases}$$
to find $x$ and $y$.

Clearing of fractions and reducing,

$$-2x + 3y = -1 \quad \cdots \quad (1),$$
$$262x - 233y = 931 \quad \cdots \quad (2).$$

Combining and eliminating $x$,

$$160y = 800; \quad \therefore \quad y = 5;$$

by substitution in (1), $\quad x = 8.$

10. Given $\left\{ \begin{array}{l} ax + by = c^2 \\ \dfrac{a(a + x)}{b(b + y)} = 1 \end{array} \right\}$ to find $x$ and $y$.

Clearing of fractions and reducing,

$$ax + by = c^2 \quad \cdots \quad (1),$$
$$ax - by = b^2 - a^2 \quad \cdots \quad (2);$$

by addition,

$$2ax = b^2 + c^2 - a^2; \quad \therefore \quad x = \frac{b^2 + c^2 - a^2}{2a};$$

by subtraction,

$$2by = c^2 + a^2 - b^2; \quad \therefore \quad y = \frac{a^2 + c^2 - b^2}{2b}.$$

MISCELLANEOUS EXAMPLES OF EQUATIONS OF THE FIRST, SECOND AND HIGHER DEGREES, CONTAINING BUT ONE UNKNOWN QUANTITY.

1. Given $\quad 3x^2 - 4 = 28 + x^2,\quad$ to find $x$:

transposing and reducing,

$$x^2 = 16; \quad \therefore \quad x = \pm 4$$

2. Given $\quad \dfrac{3x^2 + 5}{8} - \dfrac{x^2 + 29}{3} = 117 - 5x^2,\quad$ to find $x$;

Clearing of fractions, transposing and reducing,

$$x^2 = 25; \quad \therefore \quad x = \pm 5.$$

3. Given $\quad x^2 + ab = 5x^2$, to find $x$;

transposing and reducing,

$$x^2 = \frac{ab}{4}; \quad \therefore \quad x = \pm \frac{1}{2}\sqrt{ab}.$$

4. Given $\quad \dfrac{x+7}{x^2-7x} - \dfrac{x-7}{x^2+7x} = \dfrac{7}{x^2-73}$, to find $x$.

Clearing of fractions,

$$x^4 + 14x^3 - 24x^2 - 1022x - 3577 - x^4 + 14x^3 + 24x^2 - 1022x$$
$$+ 3577 = 7x^3 - 343x;$$

transposing and reducing,

$$21x^3 = 1701x, \text{ or } x^2 = 81; \quad \therefore \quad x = \pm 9.$$

5. Given $\quad \sqrt{\dfrac{x-2}{x+2}} + \sqrt{\dfrac{x+2}{x-2}} = 4$, to find $x$;

multiplying both members by $\sqrt{x+2}$,

$$\sqrt{x-2} + \frac{x+2}{\sqrt{x-2}} = 4\sqrt{x+2};$$

multiplying both members by $\sqrt{x-2}$,

$$x - 2 + x + 2 = 4\sqrt{x^2-4}, \text{ or } x = 2\sqrt{x^2-4};$$

squaring both members,

$$x^2 = 4x^2 - 16, \text{ or } x^2 = \frac{16}{3} = \frac{16}{9} \times 3; \quad \therefore \quad x = \pm \frac{4}{3}\sqrt{3}.$$

6. Given $\quad x + \sqrt{5x+10} = 8$, to find $x$;

Transposing and squaring both members,

$$5x + 10 = 64 - 16x + x^2;$$

whence, $\quad x^2 - 21x = -54;$

by the rule, $\quad x = \dfrac{21}{2} \pm \sqrt{54 + \dfrac{441}{4}} = \dfrac{21}{2} \pm \dfrac{15}{2};$

$$\therefore \quad x = 3, \quad x = 18.$$

7. Given $\quad 5\sqrt[3]{x^4} + 7\sqrt[3]{x^2} = 108,\quad$ to find $x$: make $\quad \sqrt[3]{x^2} = y;\quad$ whence, $\sqrt[3]{x^4} = y^2;$ substituting and reducing,

$$y^2 + \dfrac{7}{5}y = \dfrac{108}{5};$$

whence, $\quad y = -\dfrac{7}{10} \pm \sqrt{\dfrac{108}{5} + \dfrac{49}{100}} = -\dfrac{7}{10} \pm \dfrac{47}{10};$

$$\therefore \quad y = 4 \quad \text{and} \quad y = -\dfrac{27}{5};$$

from which, $\quad x = \pm \sqrt{y^3} = \pm 8\quad$ and $\quad x = \pm \sqrt{y^3} = \pm \sqrt{\left(-\dfrac{27}{5}\right)^3}$

8. Given $\quad 3x^2 + 10x = 57,\quad$ to find $x$.

By division,

$$x^2 + \dfrac{10}{3}x = 19; \quad \text{whence,}$$

$$x = -\dfrac{5}{3} \pm \sqrt{19 + \dfrac{25}{9}} = -\dfrac{5}{3} \pm \dfrac{14}{3};$$

$$\therefore \quad x = 3 \quad \text{and} \quad x = -6\tfrac{1}{3}.$$

9. Given $\quad (x-1)(x-2) = 1,\quad$ to find $x$.

Performing indicated operations and reducing,

$$x^2 - 3x = -1;$$

APPENDIX.

$$\therefore\ x = \frac{3}{2} \pm \sqrt{-1 + \frac{9}{4}} = \frac{3 \pm \sqrt{5}}{2} = \frac{1}{2}(3 \pm \sqrt{5}).$$

10. Given $\quad \dfrac{1}{2}x^2 - \dfrac{1}{3}x = \dfrac{5}{8}, \quad$ to find $x$.

Dividing both members by $\dfrac{1}{2}$, or multiplying by 2,

$$x^2 - \frac{2}{3}x = \frac{5}{4}; \quad \text{whence,}$$

$$x = \frac{1}{3} \pm \sqrt{\frac{5}{4} + \frac{1}{9}} = \frac{1}{3} \pm \frac{7}{6};$$

$$\therefore\ x = 1\tfrac{1}{2}, \quad x = -\frac{5}{4}.$$

11. Given $\quad \dfrac{2x-10}{8-x} - \dfrac{x+3}{x-2} = 2, \quad$ to find $x$.

Clearing of fractions,

$$2x^2 - 14x + 20 - (5x + 24 - x^2) = 20x - 32 - 2x^2;$$

reducing, $\quad x^2 - \dfrac{39}{5}x = -\dfrac{28}{5}; \quad$ whence,

$$x = \frac{39}{10} \pm \sqrt{-\frac{28}{5} + \frac{1521}{100}} = \frac{39 \pm 31}{10};$$

$$\therefore\ x = 7, \quad x = \frac{4}{5}.$$

12. Given $\quad \dfrac{1}{x-1} - \dfrac{1}{x+3} = \dfrac{1}{35}, \quad$ to find $x$.

Clearing of fractions,

$$35(x + 3 - x + 1) = x^2 + 2x - 3;$$

reducing, $\quad x^2 + 2x = 143; \quad$ whence,

$$x = -1 \pm \sqrt{144} = -1 \pm 12;$$

$$\therefore\ x = 11,\quad x = -13.$$

13. Given $x + \dfrac{24}{x-1} = 3x - 4$, to find $x$.

Clearing of fractions,
$$x^2 - x + 24 = 3x^2 - 3x - 4x + 4;$$
reducing, $\quad x^2 - 3x = 10;\quad$ whence,
$$x = \frac{3}{2} \pm \sqrt{10 + \frac{9}{4}} = \frac{3 \pm 7}{2};$$
$$\therefore\ x = 5,\quad x = -2.$$

14. Given $\dfrac{x}{x+1} + \dfrac{x+1}{x} = \dfrac{13}{6}$, to find $x$.

Clearing of fractions,
$$6x^2 + 6x^2 + 12x + 6 = 13x^2 + 13x;$$
reducing, $\quad x^2 + x = 6;\quad$ whence,
$$x = -\frac{1}{2} \pm \sqrt{6 + \frac{1}{4}} = \frac{-1 \pm 5}{2};$$
$$\therefore\ x = 2,\quad x = -3.$$

15. Given $\dfrac{x-4}{2+\sqrt{x}} = x - 8$, to find $x$.

Since $\quad x - 4 = (\sqrt{x} - 2)(\sqrt{x} + 2),$

we have, by performing indicated operations,
$$\sqrt{x} - 2 = x - 8,\quad \text{or}\quad \sqrt{x} = x - 6;$$
squaring both members,
$$x = x^2 - 12x + 36;$$
or, $\quad x^2 - 13x = -36;\quad$ whence,
$$x = \frac{13}{2} \pm \sqrt{-36 + \frac{169}{4}} = \frac{13}{2} \pm \frac{5}{2};$$

$$\therefore\ x=9,\quad x=4.$$

16. Given $\quad 17x^2 + 19x - 1848 = 0,\ $ to find $x$.

Reducing, $\quad x^2 + \dfrac{19}{17}x = \dfrac{1848}{17};\ $ whence,

$$x = -\dfrac{19}{34} \pm \sqrt{\dfrac{1848}{17} + \dfrac{361}{1156}} = \dfrac{-19 \pm 355}{34};$$

$$\therefore\ x = 9\tfrac{15}{17},\ \text{and}\ x = -11.$$

17. Given $\quad \dfrac{1}{3}x^2 + \dfrac{5}{2}x = 27,\ $ to find $x$.

Multiplying both members by 3,

$$x^2 + \dfrac{15}{2}x = 81;\ \text{whence,}$$

$$x = -\dfrac{15}{4} \pm \sqrt{81 + \dfrac{225}{16}} = \dfrac{-15 \pm 39}{4};$$

$$\therefore\ x = 6,\ \text{and}\ x = -13\tfrac{1}{2}.$$

18. Given $\quad x + 4 + \sqrt{\dfrac{x+4}{x-4}} = \dfrac{12}{x-4},\ $ to find $x$.

Transposing and reducing,

$$\sqrt{\dfrac{x+4}{x-4}} = \dfrac{28-x^2}{x-4};\ \text{or,}\ \sqrt{x+4} = \dfrac{28-x^2}{\sqrt{x-4}};$$

squaring both members and clearing of fractions,

$$x^2 - 16 = 784 - 56x^2 + x^4;$$

reducing, $\quad x^4 - 57x^2 = -800;$

by Rule, Art. 124,

$$x = \pm \sqrt{\frac{57}{2} \pm \sqrt{-800 + \frac{3249}{4}}} = \pm \sqrt{\frac{57}{2} \pm \frac{7}{2}};$$

hence, $x = \pm 5$, and $x = \pm 4\sqrt{2}$.

19. Given $\dfrac{2x + 9}{9} + \dfrac{4x - 3}{4x + 3} = 3 + \dfrac{3x - 16}{18}$, to find $x$.

Clearing of fractions,

$16x^2 + 84x + 54 + 72x - 54 = 216x + 162 + 12x^2 - 55x - 48;$

reducing, $x^2 - \dfrac{5}{4} x = \dfrac{114}{4}$; whence,

$$x = \frac{5}{8} \pm \sqrt{\frac{114}{4} + \frac{25}{64}} = \frac{5 \pm 43}{8};$$

∴ $x = 6$, and $x = -4\tfrac{3}{4}$.

20. Given $x^2 + x + 2\sqrt{x^2 + x + 4} = 20$, to find $x$.

Making $x^2 + x = y$, and reducing,

$$2\sqrt{y + 4} = 20 - y;$$

squaring both members,

$$4y + 16 = 400 - 40y + y^2;$$

reducing, $y^2 - 44y = -384$; whence,

$$y = 22 \pm \sqrt{-384 + 484} = 22 \pm 10;$$

∴ $y = 32$, and $y = 12$:

taking the first value of $y$ and substituting in the equation,

$$x^2 + x = y,$$
$$x^2 + x = 32; \text{ whence.}$$

$$x = -\frac{1}{2} \pm \sqrt{32 + \frac{1}{4}} = \frac{-1 \pm \sqrt{129}}{2};$$

taking the second value of $y$,

$$x^2 + x = 12; \text{ whence,}$$

$$x = -\frac{1}{2} \pm \sqrt{12 + \frac{1}{4}} = \frac{-1 \pm 7}{2}; \quad \therefore \ x = 3, \ x = -4.$$

21. Given $\quad \sqrt{x - \frac{1}{x}} + \sqrt{1 - \frac{1}{x}} = x$, to find $x$.

transposing,

$$\sqrt{x - \frac{1}{x}} = x - \sqrt{1 - \frac{1}{x}};$$

squaring both members,

$$x - \frac{1}{x} = x^2 - 2\sqrt{x^2 - x} + 1 - \frac{1}{x};$$

whence, by reduction,

$$x^2 - x + 1 = 2\sqrt{x^2 - x}.$$

Placing $\quad x^2 - x = y \ \cdot \ \cdot \ \cdot \ \cdot \ (1),$

$$y + 1 = 2\sqrt{y};$$

squaring both members,

$$y^2 + 2y + 1 = 4y; \text{ whence,}$$

$y^2 - 2y = -1; \quad \therefore \ y = 1 \pm \sqrt{-1 + 1}, \text{ or } y = 1;$

substituting in (1), $\quad x^2 - x = 1;$

$$\therefore \ x = \frac{1}{2} \pm \sqrt{1 + \frac{1}{4}} = \frac{1 \pm \sqrt{5}}{2}.$$

22. Given $\quad x^2 - 6x = 6x + 28$, to find $x$:

transposing, $\quad x^2 - 12x = 28;$ whence,

$$x = 6 \pm \sqrt{28 + 36} = 6 \pm 8; \quad \therefore \ x = 14, \ x = -2.$$

# ADDITIONAL EXAMPLES.

**23.** Given $\quad x^{4n} - 2x^{3n} + x^n - 6 = 0$, to find $x$:

making, $\quad x^n = y$; whence,

$$y^4 - 2y^3 + y - 6 = 0;$$

causing the second term to disappear (Arts. 263 and 313),

$$z^4 - \frac{3}{2}z = \frac{91}{16}.$$

By the rule for solving trinomial equations (Art. 124).

$$z = \sqrt{\frac{3}{4} \pm \sqrt{\frac{91}{16} + \frac{9}{16}}} = \sqrt{\frac{3 \pm 10}{4}} = \frac{1}{2}\sqrt{3 \pm 10};$$

$$\therefore \; z = \pm \frac{1}{2}\sqrt{13}, \; \text{and} \; z = \pm \frac{1}{2}\sqrt{-7};$$

but, $\quad y = \frac{1}{2} + z; \quad \therefore \; y = \frac{1 \pm \sqrt{13}}{2}, \; \text{and} \; y = \frac{1 \pm \sqrt{-7}}{2};$

also, $\quad x = \sqrt[n]{y}; \quad \therefore \; x = \sqrt[n]{\frac{1 \pm \sqrt{13}}{2}}, \; x = \sqrt[n]{\frac{1 \pm \sqrt{-7}}{2}}$

**24.** Given $\quad \dfrac{x^4 + 2x^3 + 8}{x^2 + x - 6} = x^2 + x + 8$, to find $x$.

Clearing of fractions,

$$x^4 + 2x^3 + 8 = x^4 + 2x^3 + 3x^2 + 2x - 48;$$

reducing, $\quad x^2 + \dfrac{2}{3}x = \dfrac{56}{3}$; whence,

$$x = -\frac{1}{3} \pm \sqrt{\frac{56}{3} + \frac{1}{9}} = \frac{-1 \pm 13}{3};$$

$$\therefore \; x = 4, \; \text{and} \; x = -4\tfrac{2}{3}.$$

**25.** Given $\quad x^2 - 1 = 2 + \dfrac{2}{x}$, to find $x$.

Reducing, $\quad x^3 - 3x - 2 = 0 \;\; \cdot \;\; \cdot \;\; \cdot \;\; (1);$

comparing with $\quad x^3 + px + q = 0$,
$$p = -3, \quad q = -2;$$
by Cardan's formula,
$$x = \sqrt[3]{1} + \sqrt[3]{1} = 2:$$
dividing both members of (1) by $x - 2$,
$$x^2 + 2x + 1 = 0; \quad \text{whence,}$$
$$x = -1,$$
and the two roots are each equal to $-1$; hence, the three roots are, $+2$, $-1$, and $-1$.

26. Given $\quad 2x^2 + 34 = 20x + 2$, to find $x$.

Transposing and reducing,
$$x^2 - 10x = -16; \quad \therefore \quad x = 8, \quad x = 2.$$

27. Given $\quad x^{\frac{5}{2}} = 56x^{-\frac{3}{2}} + x^{\frac{5}{6}}$, to find $x$.

multiplying both members by $x^{\frac{3}{2}}$, and reducing,
$$x^{\frac{3}{2}} - x^{\frac{3}{2}} = 56:$$

comparing with trinomial equations (Art. 124), we find
$$n = \frac{3}{2}, \quad \text{and} \quad \frac{1}{n} = \frac{2}{3};$$
hence, by rule,
$$x = \sqrt[\frac{3}{2}]{\frac{1}{2} \pm \sqrt{56\tfrac{1}{4}}} = \left(\frac{1}{2} \pm \frac{15}{2}\right)^{\frac{2}{3}};$$

taking the upper sign, $\quad x = 8^{\frac{2}{3}} = 4$,

"  lower  " $\quad x = (-7)^{\frac{2}{3}} = \sqrt[3]{49}$.

28. Given $\quad x^3 - 12x^2 + 4x + 207 = 0$, to find $x$.

ADDITIONAL EXAMPLES. 165

A superior limit of the positive roots is 13 (Art. 279); a superior limit of the negative roots (numerically), is $-7$ (Art. 281).

By the method of Art. 285, rejecting $+1$ and $-1$, which are not roots, we find

Divisors,  9,   3,   $-3$
          23,  69,  $-69$
          27,  73,  $-65$
           3,
          $-9$,
          $-1$,
           0;

hence, 9 is a commensurable root.

Dividing both members by $x - 9$, we find

$$x^2 - 3x - 23 = 0; \quad \text{or,} \quad x^2 - 3x = 23;$$

$$\therefore \; x = \frac{3}{2} \pm \sqrt{23 + \frac{9}{4}} = \frac{3 \pm \sqrt{101}}{2}.$$

29. Given $\quad x^3 + 3x^2 - 6x - 8 = 0$, to find $x$.

This is solved in a manner similar to the preceding.

A superior limit of positive roots is 4, and of negative roots (numerically), $-7$.

Divisors   4,    2,    1,   $-1$,  $-2$,  $-4$,
          $-2$, $-4$, $-8$, $+8$,  $+4$,  $+2$,
          $-8$, $-10$, $-14$, $+2$, $-2$, $-4$,
          $-2$, $-5$, $-14$, $-2$, $+1$,  $+1$,
          $+1$, $-2$, $-11$, $+1$, $+4$,  $+4$,
           ...  $-1$, $-11$, $-1$, $-2$, $-1$,
           ...   0,   $-10$,   0,  $-1$,   0;

hence,   $x = 2$,  $x = -1$,  $x = -4$.

10

**30.** Given $x^3 + 9x - 1430 = 0$, to find $x$.

By the same rule as before, we find 14 for a superior limit of the real positive roots, and from Art. 283 we see that the equation has no real negative roots. By the rule (Art. 285), we have

| 13, | 11, | 10, | 5, | 2, | 1, |
|---|---|---|---|---|---|
| $-110$, | $-130$, | $-143$, | $-286$, | $-715$, | $-1430$, |
| $-101$, | $-121$, | $-134$, | $-277$, | $-706$, | $-1421$, |
| $-8$, | $-11$, | ..., | ..., | $-353$, | $-1421$, |
| ..., | $-1$, | ..., | ..., | ..., | $-1421$, |
| ..., | 0, | ..., | ..., | ..., | $-1420$, |

hence, 11 is the only commensurable root.

Dividing both members by $x - 11$, we have

$$x^2 + 11x = -130;$$

$$\therefore x = -\frac{11}{2} \pm \sqrt{-130 + \frac{121}{4}} = \frac{-1 \pm \sqrt{-399}}{2}$$

**31.** Given $\sqrt{x} + \sqrt{x+7} = \dfrac{28}{\sqrt{x+7}}$, to find $x$.

Clearing of fractions and transposing,

$$\sqrt{x^2 + 7x} = 21 - x;$$

squaring both members and reducing,

$$49x = 441; \quad \therefore x = 9.$$

**32.** Given $\sqrt{a+x} - \sqrt{a-x} = \sqrt{ax}$, to find $x$.

Squaring both members and reducing,

$$2a - ax = 2\sqrt{a^2 - x^2};$$

squaring both members,

$$4a^2 - 4a^2 x + a^2 x^2 = 4a^2 - 4x^2;$$

reducing and dividing both members by $x$,

$$x(a^2 + 4) = 4a^2; \quad \therefore \quad x = \frac{4a^2}{a^2 + 4}.$$

33. Given $\quad \sqrt{4a + x} = 2\sqrt{b + x} - \sqrt{x}, \quad$ to find $x$.

Squaring both members and reducing,

$$(a - b) - x = -\sqrt{bx + x^2};$$

squaring both members,

$$(a - b)^2 - 2(a - b)x + x^2 = bx + x^2;$$

hence, $\quad (2a - b)x = (a - b)^2; \quad \therefore \quad x = \frac{(a - b)^2}{2a - b}.$

34. Given $\quad \sqrt{4a + x} + \sqrt{a + x} = 2\sqrt{x - 2a}, \quad$ to find $x$.

Squaring both members and reducing,

$$2\sqrt{4a^2 + 5ax + x^2} = 2x - 13a;$$

squaring both members,

$$16a^2 + 20ax + 4x^2 = 4x^2 - 52ax + 169a^2;$$

reducing, $\quad 72ax = 153a^2; \quad \therefore \quad x = \frac{17a}{8}.$

35. Given $\quad \dfrac{1 + x^3}{(1 + x)^2} + \dfrac{1 - x^3}{(1 - x)^2} = a, \quad$ to find $x$.

Dividing both terms of the first fraction by $(1 + x)$, and of the second by $1 - x$, we have

$$\frac{1 - x + x^2}{1 + x} + \frac{1 + x + x^2}{1 - x} = a;$$

clearing of fractions,

$$1 - 2x + 2x^2 - x^3 + 1 + 2x + 2x^2 + x^3 = a - ax^2;$$

reducing, $\quad (4 + a)x^2 = a - 2; \quad \therefore \quad x = \pm \sqrt{\dfrac{a - 2}{a + 4}}.$

## MISCELLANEOUS EXAMPLES OF SIMULTANEOUS EQUATIONS OF THE SECOND AND HIGHER DEGREES.

1. Given $\begin{cases} x^3 + y^3 = 18xy & \ldots \quad (1) \\ x + y = 12 & \ldots \quad (2) \end{cases}$ to find $x$ and $y$.

Make $\quad x = v + w$, and $y = v - w$.

From (2), we have $(v + w) + (v - w) = 12$; $\quad \therefore v = 6$.

From (1), we have, $(v + w)^3 + (v - w)^3 = 18(v^2 - w^2)$;

or, reducing, $\quad v^3 + 3vw^2 = 9(v^2 - w^2)$;

substituting the value of $v$, we have

$216 + 18w^2 = 9(36 - w^2)$, or $27w^2 = 108$; $\quad \therefore w = \pm 2$,

hence,

$x = v + w = 6 \pm 2 = 8$ and $4$; $y = v - w = 6 \mp 2 = 4$ and $8$.

2. Given $\begin{cases} x^2 + y^2 = 53 & \ldots \quad (1) \\ xy = 14 & \ldots \quad (2) \end{cases}$ to find $x$ and $y$.

Multiplying both members of (2) by 2, and adding and subtracting, we have

$x^2 + 2xy + y^2 = 81$; $\quad \therefore x + y = \pm 9$,

$x^2 - 2xy + y^2 = 25$; $\quad \therefore x - y = \pm 5$;

hence, $\quad x = +7$, and $-7$ $\quad y = +2$, and $-7$.

3. Given $\begin{cases} x^4 + y^4 = 82 & \ldots \quad (1) \\ x + y = 4 & \ldots \quad (2) \end{cases}$ to find $x$ and $y$.

Raising both members of (2) to the 4th power, adding to (1), member to member, and dividing by 2,

$x^4 + 2x^3y + 3x^2y^2 + 2xy^3 + y^4 = 169$;

extracting the square root of both members,

$$x^2 + xy + y^2 = 13 \quad . \quad . \quad . \quad (3);$$

squaring both members of (1),

$$x^2 + 2xy + y^2 = 16 \quad . \quad . \quad . \quad (4);$$

subtracting (3) from (4), member from member,

$$xy = 3, \quad \text{or} \quad 3xy = 9 \quad . \quad . \quad (5);$$

subtracting from (3), member from member,

$$x^2 - 2xy + y^2 = 4;$$

whence, $\quad x - y = \pm 2 \quad . \quad . \quad . \quad (6).$

Combining (2) and (6),

$$x = 3, \quad \text{and} \quad 1; \quad y = 1, \quad \text{and} \quad 3.$$

4. Given $\quad \begin{cases} 5x + 3y = 19 \quad . \quad . \quad (1) \\ 7x^2 - 2y^2 = 10 \quad . \quad . \quad (2) \end{cases}$ to find $x$ and $y$

From (1), we find

$$x = \frac{19 - 3y}{5}; \quad \therefore \quad x^2 = \frac{361 - 114y + 9y^2}{25}.$$

Substituting in (2), and reducing,

$$y^2 - \frac{798}{13} = -\frac{2277}{13};$$

whence, $\quad y = \frac{399}{13} \pm \sqrt{\frac{-2277}{13} + \frac{159201}{169}} = \frac{399 \pm 360}{13};$

hence, $\quad y = \frac{759}{13}, \quad \text{and} \quad y = 3;$

by substitution, $\quad x = -\frac{406}{13}, \quad \text{and} \quad x = 2.$

5. Given $\begin{cases} x + 4y = 14 \quad . \quad . \quad (1) \\ y^2 + 4x = 2y + 11 \quad (2) \end{cases}$ to find $x$ and $y$.

From (1), $x = 14 - 4y$, or $4x = 56 - 16y$;

subtracting and reducing,

$$y^2 - 18y = -45;$$

whence, $y = 9 \pm \sqrt{-45 + 81} = 9 \pm 6 = 15$ and $3$;

hence, $x = 2$ and $-46$.

6. Given $\begin{cases} x^2 + 4y^2 = 256 - 4xy \quad . \quad . \quad (1) \\ 3y^2 - x^2 = 39 \quad . \quad . \quad . \quad (2) \end{cases}$ to find $x$ and $y$.

Transposing in (1), and extracting the square root of both members,

$$x + 2y = \pm 16; \quad \therefore \quad x = \pm 16 - 2y;$$

or, $\qquad x^2 = 256 \mp 64y + 4y^2;$

substituting in (2), and reducing,

$$y^2 \mp 64y = -295;$$

whence, $y = \pm 32 \pm \sqrt{-295 + 1024} = \pm 32 \pm 27;$

$\therefore \quad y = \pm 59$ and $\pm 5;$

substituting, $x = \pm 102$ and $\pm 6$.

7. Given $\begin{cases} x^2 - y^2 = 24 \quad . \quad . \quad (1) \\ x^2 + xy = 84 \quad . \quad . \quad (2) \end{cases}$ to find $x$ and $y$.

Subtracting (1) from (2), member from member,

$$y^2 + xy = 60 \quad . \quad . \quad . \quad (3);$$

adding (3) to (2), member to member,

$$x^2 + 2xy + y^2 = 144; \quad \therefore \quad x + y = \pm 12 \quad . \quad . \quad . \quad (4).$$

Dividing (1) by (4), member by member,

$$x - y = \pm 2;$$

hence, $\quad x = \pm 7, \quad y = \pm 5.$

8. Given $\quad \begin{cases} x - y = 4 \quad \ldots \quad (1) \\ xy = 45 \quad \ldots \quad (2) \end{cases}$ to find $x$

From (1), $\quad x = y + 4,\quad$ which, in (2), gives

$$y^2 + 4y = 45; \quad \therefore \ y = -2 \pm \sqrt{45 + 4} = -2 \pm 7;$$

$$\therefore \ y = +5, \quad \text{and} \quad -9;$$

and by substitution, $\quad x = +9, \quad \text{and} \quad -4.$

9. Given $\quad \begin{cases} xy + xy^2 = 12 \quad \ldots \quad (1) \\ x + xy^3 = 18 \quad \ldots \quad (2) \end{cases}$ to find $x$ and $y$.

Dividing (2) by (1), member by member, and reducing,

$$\frac{1+y^3}{y(1+y)} = \frac{3}{2}, \quad \text{or} \quad \frac{1-y+y^2}{y} = \frac{3}{2};$$

by reduction, $\quad y^2 - \frac{5}{2}y = -1;$

whence, $\quad y = \frac{5}{4} \pm \sqrt{-1 + \frac{25}{16}} = \frac{5}{4} \pm \frac{3}{4};$

$$\therefore \ y = 2, \quad \text{and} \quad y = \frac{1}{x};$$

by substitution, $\quad x = 2, \quad \text{and} \quad x = 16.$

10. Given $\quad \begin{cases} \dfrac{1}{x} \cdot \dfrac{1}{y} = a \quad \ldots \quad (1) \\ \dfrac{1}{x^2} + \dfrac{1}{y^2} = b \quad \ldots \quad (2) \end{cases}$ to find $x$ and $y$.

From (1), we find,
$$\frac{1}{y} = a - \frac{1}{x}, \text{ or } \frac{1}{y^2} = a^2 - \frac{2a}{x} + \frac{1}{x^2};$$

substituting in (2), and reducing,
$$\left(\frac{1}{x}\right)^2 - a\left(\frac{1}{x}\right) = \frac{b - a^2}{2};$$

$$\therefore \frac{1}{x} = \frac{a}{2} \pm \sqrt{\frac{b - a^2}{2} + \frac{a^2}{4}} = \frac{a \pm \sqrt{2b - a^2}}{2};$$

by substitution,
$$\frac{1}{y} = \frac{a \mp \sqrt{2b - a^2}}{2};$$

hence,
$$x = \frac{2}{a \pm \sqrt{2b - a^2}}, \text{ and } y = \frac{2}{a \mp \sqrt{2b - a^2}}.$$

11. Given
$$\begin{cases} \dfrac{x^2}{y^2} + \dfrac{4x}{y} = \dfrac{85}{9} & \cdots \text{ (1)} \\ x - y = 2 & \cdots \text{ (2)} \end{cases}$$
to find $x$ and $y$.

Clearing (1) of fractions,
$$9x^2 + 36xy = 85y^2 \cdots \text{ (3)}.$$

From (2), $x = 2 + y$; $\therefore x^2 = 4 + 4y + y^2$;

substituting in (3), and reducing,
$$y^2 - \frac{27}{10}y = \frac{9}{10};$$

whence,
$$y = \frac{27}{20} \pm \sqrt{\frac{9}{10} + \frac{729}{400}} = \frac{27 \pm 33}{20};$$

$$\therefore y = 3 \text{ and } y = -\frac{3}{10}$$

by substitution, $x = 5$ and $x = \dfrac{17}{10}.$

ADDITIONAL EXAMPLES.   173

12. Given $\begin{cases} \dfrac{x^2}{y^2}+\dfrac{y^2}{x^2}+\dfrac{x}{y}+\dfrac{y}{x}=\dfrac{27}{4} \cdots (1) \\ x-y=2 \cdots\cdots\cdots (2) \end{cases}$ to find $x$ and $y$.

Make $\quad \dfrac{x}{y}+\dfrac{y}{x}=z \quad (3);\quad$ whence, by squaring, &c.,

$$\dfrac{x^2}{y^2}+\dfrac{y^2}{x^2}=z^2-2;$$

hence, from (1), by substitution and reduction,

$$z^2+z=\dfrac{35}{4};\quad\therefore\ z=-\dfrac{1}{2}\pm\sqrt{\dfrac{35}{4}+\dfrac{1}{4}}=\dfrac{-1\pm 6}{2};$$

or, $\qquad\qquad\qquad z=\dfrac{5}{2}\ \text{ and }\ -\dfrac{7}{4};$

substituting the positive value of $z$ in (3), and clearing of fractions,

$$x^2+y^2=\dfrac{5}{2}xy\ \cdot\ \cdot\ \cdot\ (4),$$

From (2), $\quad x=y+2;\quad \therefore\ x^2=y^2+4y+4,$

and $\qquad\qquad\qquad xy=y^2+2y;$

substituting in (4) and reducing,

$$y^2+2y=8;\quad \therefore\ y=-1\pm\sqrt{8+1}=-1\pm 3;$$

hence, $\qquad y=2\ \text{ and }\ y=-4;$

by substitution, $\quad x=4\ \text{ and }\ y=-2.$

13. Given $\begin{cases} x^2+y^2+z^2=84 \ \cdot\ \cdot\ \cdot\ (1) \\ x+y+z=14 \ \cdot\ \cdot\ \cdot\ (2) \\ xz=y^2 \qquad\qquad \cdot\ \cdot\ \cdot\ (3) \end{cases}$ to find $x$, $y$, and $z$.

Substituting in (1) and (2) the value of $y$ from (3), and reducing

$$x^2+2xz+z^2=84+xz\quad\text{or}\quad x+z=\sqrt{84+xz}\ \cdot\ \cdot\ (4).$$

From (2) and (3), $\qquad\qquad x+z=14-\sqrt{xz}\ \cdot\ \cdot\ (5)$

Equating the second members,
$$14 - \sqrt{xz} = \sqrt{84 + xz};$$
squaring both members,
$$196 - 28\sqrt{xz} + xz = 84 + xz;$$
reducing, $\quad\quad\sqrt{xz} = 4, \text{ or } xz = 16 \cdot\cdot\cdot\text{ (6)};$

hence, from (5), $\quad\quad\quad x + z = 10 \cdot\cdot\cdot\text{ (7)};$

substituting in (7) for $z$ its value $\dfrac{16}{x}$, and reducing,
$$x^2 - 10x = -16;$$
whence, $\quad\quad x = 5 \pm \sqrt{-16 + 25} = 5 \pm 3;$

or, $\quad\quad\quad\quad x = 8, \quad x = 2:$

by substitution, $z = 2, \quad z = 8,$ and $y = \pm 4.$

14. Given $\begin{cases} x + y + \sqrt{x+y} = 12 \cdot\cdot\text{ (1)} \\ x^2 + y^2 = 41 \cdot\cdot\text{ (2)} \end{cases}$ to find $x$ and $y$.

From (1), by transposition,
$$\sqrt{x+y} = 12 - (x+y);$$
squaring both members,
$$x + y = 144 - 24(x+y) + (x+y)^2;$$
reducing, $\quad\quad (x+y)^2 - 25(x+y) = -144;$
$$\therefore\; x + y = +\frac{25}{2} \pm \sqrt{-144 + \frac{625}{4}} = +\frac{25 \pm 7}{2};$$
whence, $\quad\quad x + y = 16,$ or $x + y = 9.$

The first value does not satisfy (1), unless the radical have the negative sign; adopting, therefore, the second value, from which
$$x = 9 - y, \text{ or } x^2 = 81 - 18y + y^2,$$

which in (2) gives, after reduction,

$$y^2 - 9y = -20; \quad \text{whence,}$$

$$y = \frac{9}{2} \pm \sqrt{-20 + \frac{81}{4}} = \frac{9 \pm 1}{2}; \quad \therefore \ y = 5 \ \text{and} \ y = 4;$$

by substitution, $\quad x = 4 \ \text{and} \ x = 5.$

15. Given $\quad \begin{cases} x^3 - y^3 = 117 \ \cdots \ (1) \\ x - y = 3 \ \cdots \ (2) \end{cases}$ to find $x$ and $y$.

Cubing both members of (2), subtracting from (1), member from member, and dividing both members by 3, we have

$$x^2y - xy^2 = 30, \quad \text{or} \quad (x - y)xy = 30 \ \cdots \ (3);$$

dividing (3) by (2), member by member,

$$xy = 10; \quad \therefore \ y = \frac{10}{x};$$

substituting in (2), and reducing,

$$x^2 - 3x = 10;$$

whence, $\quad x = \dfrac{3}{2} \pm \sqrt{10 + \dfrac{9}{4}} = \dfrac{3}{2} \pm \dfrac{7}{2};$

$$\therefore \ x = 5, \quad x = -2;$$

by substitution, $\quad y = 2, \ \text{and} \ y = -5.$

16. Given $\quad \begin{cases} x^2 + x\sqrt[3]{xy^2} = 208 \ \cdot \ \cdot \ (1) \\ y^2 + y\sqrt[3]{x^2y} = 1053 \ \cdot \ \cdot \ (2) \end{cases}$ to find $x$ and $y$.

These equations may be written,

$$x^{\frac{4}{3}} + x^{\frac{4}{3}}y^{\frac{2}{3}} = 208, \quad \text{or} \quad x^{\frac{4}{3}}(x^{\frac{2}{3}} + y^{\frac{2}{3}}) = 208 \ \cdots \ (3),$$

$$y^{\frac{4}{3}} + y^{\frac{4}{3}}x^{\frac{2}{3}} = 1053, \quad y^{\frac{4}{3}}(y^{\frac{2}{3}} + x^{\frac{2}{3}}) = 1053 \ \cdots \ (4)$$

Dividing (3) by (4), member by member,

$$\frac{x^{\frac{4}{3}}}{y^{\frac{4}{3}}} = \frac{208}{1053} = \frac{16}{81} \quad \cdots \quad (5);$$

extracting the 4th root of both members,

$$\frac{x^{\frac{1}{3}}}{y^{\frac{1}{3}}} = \frac{2}{3}; \quad \therefore \quad x^{\frac{1}{3}} = \frac{2y^{\frac{1}{3}}}{3};$$

substituting in (3),

$$\frac{64}{729} y^2 + \frac{16}{81} y^2 = 208; \quad \text{whence,}$$

$$\frac{208}{729} y^2 = 208, \quad \text{or} \quad y^2 = 729; \quad \therefore \quad y = \pm 27,$$

and by substitution, $\qquad\qquad\qquad\qquad\qquad x = \pm\ 8.$

## MISCELLANEOUS PROBLEMS.

1. A courier starts from a place and travels at the rate of 4 miles per hour; a second courier starts after him, an hour and a half later, and travels at the rate of 5 miles per hour: in how long a time will the second overtake the first, and how far will he travel?

Let $x$ denote the number of hours travelled by 2d courier:
then will $x + 1\frac{1}{2}$ "     "     "     "     1st "
         $5x$     "     "     " miles    "     2d "
and    $4(x + 1\frac{1}{2})$ "     "     "     "     "     1st "

From the conditions of the problem,

$$5x = 4(x + 1\tfrac{1}{2}); \quad \therefore \quad x = 6 \quad \text{and} \quad 5x = 30.$$

2. A person buys 4 houses for $8000; for the second he gave half as much again as for the first; for the third, half as much again as for the second; and for the fourth, as much as for the first and third together: what does he give for each?

Let $x$  denote the amount paid for 1st house: then will

$x + \dfrac{x}{2}$  "  "  "  "  "  2d  "

$2x + \dfrac{x}{4}$  "  "  "  "  "  3d  "

$3x + \dfrac{x}{4}$  "  "  "  "  "  4th  "

From the conditions of the problem,

$$8x = 8000; \qquad \therefore \quad x = 1000$$

$$x + \frac{x}{2} = 1500$$

$$2x + \frac{x}{4} = 2250$$

$$3x + \frac{x}{4} = 3250$$

3. A and B engaged in play: after A had lost $20, he had one third as much as B; but continuing to play, he won back his $20, together with $50 more, and he then found that he had half as much again as B: with what sums did they begin?

Let $x$ and $y$ denote the sums with which A and B began.

Then from the conditions,

$$x - 20 = \frac{y + 20}{3}$$

$$x + 50 = (y - 50) \times 1\tfrac{1}{2};$$

whence, $\quad 3x - 60 = y + 20; \quad \therefore \quad x = 70$

$\qquad 2x + 100 = 3y - 150; \qquad y = 130.$

4. A can do a piece of work in 10 days, which A and B together can do in 7 days: in how many days can B do it alone?

Let $x$ denote the number of days.

Since A and B together can do $\frac{1}{7}$ of the work in 1 day, A can do $\frac{1}{10}$ of it, and B, $\frac{1}{x}$ of it in 1 day; hence, from the relations existing,

$$\frac{1}{10} + \frac{1}{x} = \frac{1}{7}; \qquad \therefore \frac{1}{x} = \frac{3}{70}, \quad \text{or} \quad x = 23\tfrac{1}{3}.$$

5. A person has $650 invested in two parts: the first part draws interest at 3 per cent, and the second at $3\tfrac{1}{2}$ per cent, and his total income is $20 per annum: how much has he invested at each rate?

Let $x$ denote the number of dollars at 3 per cent: then will $650 - x$ denote the number at $3\tfrac{1}{2}$.

From the conditions,

$$\frac{3x}{100} + \frac{650 - x}{100} \times 3\tfrac{1}{2} = 20;$$

whence, $\qquad 3x + 2275 - 3\tfrac{1}{2}x = 2000;$

$\therefore \ \dfrac{x}{2} = 275,\ $ or $\ x = 550;\qquad \therefore\ 650 - x = 100.$

6. A boatman rows with the tide, in the channel, 18 miles in $1\tfrac{1}{2}$ hours; he rows near the shore against the tide, which is then only three-fifths as strong as in the channel, 18 miles in $2\tfrac{1}{4}$ hours: what is the velocity of the tide per hour in the channel?

Let $x$ denote the velocity of the tide in the channel:

then, $\dfrac{3x}{5}$ " " " " " near shore;

and $\left(18 - \dfrac{3x}{2}\right) \div 1\tfrac{1}{2}$ will denote the rate of rowing, neglecting tide

also, $\left(18 + \dfrac{27x}{20}\right) \div 2\tfrac{1}{4}$ " " " " " "

hence, $\qquad \left(18 - \dfrac{3x}{2}\right) \times \dfrac{2}{3} = \left(18 + \dfrac{27x}{20}\right) \times \dfrac{4}{9};$

or, $\qquad 12 - x = 8 + \dfrac{3x}{5}, \qquad \therefore\ x = 2\tfrac{1}{2}.$

7. A garrison had provisions for 30 months, but at the end of 4 months the number of troops was doubled, and 3 months afterwards it was reinforced by 400 troops more, and the provisions were exhausted in 15 months: how many troops were there in the garrison at first?

Let $x$ denote the number of men at first; then will $30x$ denote the number of months that one person could subsist on the provisions, or the number of monthly rations in the garrison.

$4x$ denotes the number of monthly rations used in 4 months,
$6x$ " " " " " the next 3 "
$(2x + 400)8$ " " " " " " 8 "

hence, $26x + 3200 = 30x$, or $4x = 3200$, $\therefore x = 800$.

8. What is the number whose square exceeds the number itself by 6?

Let $x$ denote the number.

From the conditions,
$$x^2 - x = 6; \quad \therefore x = \tfrac{1}{2} \pm \sqrt{6 + \tfrac{1}{4}} = \frac{1 \pm 5}{2};$$
$$\therefore x = 3 \text{ and } -2.$$

9. Find two numbers such that their sum shall be 15, and the sum of their squares 117.

Let $x$ and $y$ denote the numbers.

From the conditions of the problem,
$$x + y = 15 \quad \cdots \quad (1),$$
$$x^2 + y^2 = 117 \quad \cdots \quad (2).$$

From (1) $x = 15 - y$, or $x^2 = 225 - 30y + y^2$;

substituting in (2) and reducing,

$$y^2 - 15y = -54,$$

$$y = \frac{15}{2} \pm \sqrt{-54 + \frac{225}{4}} = \frac{15}{2} \pm \frac{3}{2}, \text{ or } y = 9 \text{ and } x = 6,$$

$$\text{or } x = 9 \text{ and } y = 6.$$

10. A cask whose contents is 20 gallons, is filled with brandy; a certain quantity is drawn off into another cask of the same size, after which the latter is filled with water: the first cask is then filled with this mixture; it then appears that if $6\frac{2}{3}$ gallons of this mixture be drawn from the first into the second cask, there will be equal quantities of brandy in each. How much brandy was first drawn off?

Let $x$ denote the number of gallons first drawn off. Then will $20 - x$ denote the quantity remaining as well as the quantity of water added to the second cask; $\frac{x}{20}$ will denote the quantity of brandy in each gallon of the mixture, and

$$x \times \frac{x}{20}, \text{ or } \frac{x^2}{20}$$

will denote the quantity of brandy returned to the first cask, which will, therefore, contain

$$20 - x + \frac{x^2}{20}$$

gallons of brandy. Each gallon of this new mixture will contain $\frac{1}{20}$ of the brandy in the cask, or

$$\frac{400 - 20x + x^2}{400};$$

hence. $6\frac{2}{3}$ gallons will contain

$$\frac{400 - 20x + x^2}{60}$$

gallons; and after this is drawn off, 10 gallons must remain; hence,

$$\frac{400 - 20x + x^2}{20} - \frac{400 - 20x + x^2}{60} = 10;$$

whence, $\qquad 800 - 40x + 2x^2 = 600,$

or, $\qquad x^2 - 20x = 100;$

$\therefore\ x = 10 \pm \sqrt{-100 + 100},\ \text{ or }\ x = 10.$

11. What number added to its square will produce 42?

Let $x$ denote the number.

From the conditions of the problem,

$$x^2 + x = 42;$$

$\therefore\ x = -\tfrac{1}{2} + \sqrt{42 + \tfrac{1}{4}} = \dfrac{-1 \pm 13}{2};\ \ \therefore\ x = 6 \text{ and } x = -7$

12. The difference of two numbers is 9, and their sum multiplied by the greater gives 266: what are the two numbers?

Let $x$ and $y$ denote the numbers.

From the conditions of the problem,

$$x - y = 9 \ \cdot\ \cdot\ \cdot\ (1),$$
$$x(x + y) = 266 \ \cdot\ (2).$$

From (1), $\qquad y = x - 9;\qquad$ substituting in (2),

$$x(2x - 9) = 266,\ \text{ or }\ x^2 - \frac{9}{2}x = 133;$$

whence, $\qquad x = \dfrac{9}{4} \pm \sqrt{133 + \dfrac{81}{16}} = \dfrac{9 \pm 47}{4};$

$\therefore\ x = 14,\qquad x = -9\tfrac{1}{2};$

whence, $\qquad y = 5,\qquad y = -18\tfrac{1}{2}.$

13. A person travelled 105 miles: if he had travelled 2 miles

per hour slower, he would have been 6 hours longer in completing the journey: how many miles did he travel per hour?

Let $x$ denote the number of miles travelled per hour. Then will $\dfrac{105}{x}$ denote the number of hours.

From the conditions,

$$\frac{105}{x-2} = \frac{105}{x} + 6, \quad \text{or} \quad 105x = 105x - 210 + 6x^2 - 12x;$$

reducing, $\quad x^2 - 2x = 35;$

$\therefore \quad x = 1 \pm \sqrt{35 + 1} = 1 \pm 6; \quad \therefore \quad x = 7.$

14. The continued product of four consecutive numbers is 3024: what are the numbers?

Let $x$ denote the least number.

From the conditions of the problem.

$$x(x+1)(x+2)(x+3) = 3024,$$
or $\quad x^4 + 6x^3 + 11x^2 + 6x - 3024 = 0.$

A superior limit of the real positive roots is 9 (Art. 279). Neglecting the divisor 1, and all negative divisors, we may proceed by the rule (Art. 285), as follows:

| 9, | 8, | 7, | 6, | 4, | 3, | 2, |
|---|---|---|---|---|---|---|
| − 336, | − 378, | − 432, | − 504, | − 756, | − 1008, | − 1512, |
| − 330, | − 372, | − 426, | − 498, | − 750, | − 1002, | − 1506, |
| ··, | ··, | ··, | − 83, | ··, | − 334, | − 753, |
| ··, | ··, | ··, | − 72, | ··, | − 323, | − 742, |
| ··, | ··, | ··, | − 12, | ··, | ··, | − 371, |
| ··, | ··, | ··, | − 6, | ··, | ··, | − 365, |
| ··, | ··, | ··, | − 1, | ··, | ··, | ··, |
| ··, | ··, | ··, | − 0, | ··, | ··, | ·· |

Hence, 6 is the required value of $x$, and the numbers are 6, 7, 8 and 9.

# ADDITIONAL EXAMPLES. 183

15. Two couriers start at the same instant for a point 39 miles distant; the second travels a quarter of a mile per hour faster than the first, and reaches the point one hour ahead of him: at what rates do they travel?

Let $x$ denote the number of miles per hour of first courier. Then will $\dfrac{39}{x}$ denote the number of hours he travels.

From the conditions,

$$\frac{39}{x} - 1 = \frac{39}{x + \frac{1}{4}}, \quad \text{or} \quad 39x + \frac{39}{4} - x^2 - \frac{1}{4}x = 39x;$$

reducing, $\qquad x^2 + \dfrac{1}{4}x = \dfrac{39}{4};$

$$\therefore \quad x = -\frac{1}{8} \pm \sqrt{\frac{39}{4} + \frac{1}{64}} = \frac{-1 \pm 25}{8}, \quad \text{or} \quad x = 3.$$

16. The fore-wheels of a wagon are $5\frac{1}{4}$ feet, and the hind-wheels $7\frac{1}{8}$ feet in circumference; after a certain journey, it is found that the fore-wheels have made 2000 revolutions more than the hind-wheels: how far did the wagon travel?

Let $x$ denote the number of feet.

From the conditions of the problem,

$$\frac{x}{5\frac{1}{4}} - \frac{x}{7\frac{1}{8}} = 2000;$$

multiplying both members by $\dfrac{1197}{32}$,

$$7\tfrac{1}{8}x - 5\tfrac{1}{4}x = \frac{2394000}{32} = \frac{598500}{8},$$

$$57x - 42x = 598500,$$

$$15x = 598500,$$

$$x = 39900.$$

17. A wine merchant has 2 kinds of wine; the one costs 9 shillings per gallon, and the other 5. He wishes to mix them together in such quantities that he may have 50 gallons of the mixture, and so that each gallon of the mixture shall cost 8 shillings.

Let $x$ and $y$ denote the number of gallons of each, respectively.

From the conditions,
$$x + y = 50 \quad \ldots \quad (1),$$
$$9x + 5y = 8(x + y) \quad \ldots \quad (2);$$
substituting for $x + y$ its value in (2),
$$9x + 5y = 400 \quad \ldots \quad (3);$$
combining (1) and (3),
$$4y = 50; \quad \therefore \ y = 12\tfrac{1}{2}, \text{ and } x = 37\tfrac{1}{2}.$$

18. A owes $1200 and B, $2500, but neither has enough to pay his debts. Says A to B, "Lend me the eighth part of your fortune, and I can pay my debts." Says B to A, "Lend me the ninth part of your fortune, and I can pay mine:" what fortune had each?

Let $x$ and $y$ denote the number of dollars in the fortunes of A and B.

From the conditions of the problem,
$$x + \frac{y}{8} = 1200, \quad \text{or} \quad 8x + y = 9600,$$
$$y + \frac{x}{9} = 2500, \quad \text{or} \quad x + 9y = 22500;$$
combining and eliminating $x$,
$$71y = 170400; \quad \therefore \ y = 2400, \quad x = 900.$$

19. A person has two kinds of goods, 8 pounds of the first, and 9 of the second, cost together $18,40; 20 pounds of the first, and 16 of the second, cost together $36,40: how much does each cost per pound?

Let $x$ and $y$ denote the cost of a pound of each in cents.
From the conditions of the problem,
$$8x + 9y = 1846,$$
$$20x + 16y = 3640;$$
combining and eliminating $x$,
$$13y = 1950: \quad \therefore \quad y = 150, \quad \text{and} \quad x = 62.$$

20. What fraction is that to the numerator of which if 1 be added the result will be $\frac{1}{3}$, but if 1 be added to the denominator the result will be $\frac{1}{4}$?

Let $x$ denote the numerator, and $y$ the denominator.

From the conditions of the problem,
$$\frac{x+1}{y} = \frac{1}{3}, \quad \text{or} \quad 3x + 3 = y,$$
$$\frac{x}{1+y} = \frac{1}{4}, \quad \text{or} \quad 4x = 1 + y;$$

hence, by combination, $x = 4$ and $y = 15$. *Ans.* $\frac{4}{15}$.

21. A shepherd was plundered by three parties of soldiers. The first party took $\frac{1}{4}$ of his flock and $\frac{1}{4}$ of a sheep; the second took $\frac{1}{3}$ of what remained and $\frac{1}{3}$ of a sheep; the third took $\frac{1}{2}$ of what then remained and $\frac{1}{2}$ of a sheep, which left him but 25 sheep: how many had he at first?

Let $x$ denote the number of sheep. Then, after being plundered by the 1st party, he would have
$$x - \left(\frac{x}{4} + \frac{1}{4}\right) = \frac{3x-1}{4} \quad \text{sheep};$$

after being plundered by the 2d party, he would have
$$\frac{3x-1}{4} - \left(\frac{3x-1}{12} + \frac{1}{3}\right) = \frac{x-1}{2};$$

after being plundered by the 3d party, he would have

$$\frac{x-1}{2} - \left(\frac{x-1}{4} + \frac{1}{2}\right) = \frac{x-3}{4};$$

from the conditions of the problem,

$$\frac{x-3}{4} = 25, \quad \text{or} \quad x-3 = 100; \quad \therefore \quad x = 103.$$

22. What two numbers are those whose product is 63, and the square of whose sum is equal to 64 times the square of their difference?

Let $x$ and $y$ denote the two numbers.

From the conditions of the problem,

$$xy = 63 \quad \ldots \ldots \quad (1),$$
$$(x+y)^2 = 64(x-y)^2 \quad \ldots \quad (2);$$

extracting the square root of both members of (2),

$$x + y = 8(x-y), \quad \text{or} \quad 7x = 9y; \quad \therefore \quad x = \tfrac{9}{7}y;$$

substituting in (1), $\quad \tfrac{9}{7}y^2 = 63;$

$$\therefore \quad y^2 = 49 \quad \text{and} \quad y = 7, \quad \text{also} \quad x = 9.$$

23. The sum of two numbers multiplied by the greater gives 209; their sum multiplied by their difference gives 57: what are the two numbers?

Let $x$ and $y$ denote the numbers.

From the conditions of the problem,

$$(x+y)x = 209, \quad \text{or} \quad x^2 + xy = 209 \quad \ldots \quad (1),$$
$$(x+y)(x-y) = 57, \quad \text{or} \quad x^2 - y^2 = 57 \quad \ldots \quad (2);$$

subtracting (2) from (1), member from member,

$$xy + y^2 = 152 \quad \ldots \quad (3);$$

adding (3) and (1), member to member,

$$x^2 + 2xy + y^2 = 361; \quad \therefore \quad x + y = 19;$$

hence, from (1),	$x = \dfrac{209}{19} = 11$;   also,   $y = 8$.

24. Three numbers are in arithmetical progression; their sum is 15, and the sum of their cubes is 495: what are the numbers?

Let $x$, $y$ and $z$ denote the numbers,

From the conditions of the problem,

$$y - x = z - y \quad . \quad . \quad (1)$$
$$x + y + z = 15 \quad . \quad . \quad (2)$$
$$x^3 + y^3 + z^3 = 495 \quad . \quad . \quad (3);$$

from (1),   $2y = z + x$,   which in (2), gives   $y = 5$;

substituting in (2) and (3),

$$z + x = 10 \quad . \quad . \quad (4)$$
$$z^3 + x^3 = 370 \quad . \quad . \quad (5);$$

dividing (5) by (4), member by member,

$$z^2 - zx + x^2 = 37 \quad . \quad . \quad (6);$$

squaring both members of (4),

$$z^2 + 2zx + x^2 = 100 \quad . \quad . \quad (7);$$

combining (6) and (7),

$3zx = 63$,   or   $zx = 21$;   $\therefore$   $z = \dfrac{21}{x}$;

substituting in (4),

$x + \dfrac{21}{x} = 10$,   or   $x^2 - 10x = -21$;

$\therefore$   $x = 5 \pm \sqrt{-21 + 25} = 5 \pm 2$;   hence,   $x = 7$, or 3.

25. Divide the number 16 into two parts such that 25 times the square of the first shall be equal to 9 times the square of the second.

Let $x$ denote one part; then will $16 - x$ denote the other. From the conditions,

$$25x^2 = 9(256 - 32x + x^2) = 2304 - 288x + 9x^2;$$

reducing, $\qquad x^2 + 18x = 144;$

$$\therefore \; x = -9 \pm \sqrt{144 + 81} = -9 \pm 15, \text{ or } x = 6,$$

since the negative value does not satisfy the problem understood in the numerical sense.

26. There are two numbers such that the greater multiplied by the square root of the less is 18, and the less multiplied by the square root of the greater is 12 : what are the numbers?

Let $x$ and $y$ denote the numbers.

From the conditions of the problem,

$$y\sqrt{x} = 18 \quad . \quad . \quad (1)$$
$$x\sqrt{y} = 12 \quad . \quad . \quad (2);$$

multiplying (1) by (2), member by member,

$$\sqrt{(xy)^3} = 216; \quad \therefore \; xy = 36 \quad . \quad . \quad (3);$$

adding (1) and (2), member to member,

$$(\sqrt{x} + \sqrt{y})\sqrt{xy} = 30 \quad . \quad . \quad (4),$$

or $\qquad \sqrt{x} + \sqrt{y} = 5;$

squaring both members,

$$x + y + 2\sqrt{xy} = 25 \quad . \quad . \quad (5),$$

or $\qquad\qquad x + y = 13 \quad . \quad (6);$

combining (3) and (6),

$$x = 9, \quad y = 4.$$

27. What two numbers are those the square of the greater of which being multiplied by the lesser gives 147, and the square of the lesser being multiplied by the greater gives 63?

Let $x$ and $y$ denote the numbers.

From the conditions of the problem,
$$x^2y = 147 \quad . \quad . \quad (1)$$
$$xy^2 = 63 \quad . \quad . \quad (2);$$

multiplying (1) and (2), member by member,
$$x^3y^3 = 9261 \quad . \quad . \quad (3),$$
or
$$xy = 21 \quad . \quad . \quad (4);$$

dividing (2) by (4), member by member,

$y = 3$; in like manner, $x = 7$.

This method of solution might be applied to the equations of the preceding example.

28. There are two numbers whose difference is 2, and the product of their cubes is 42875: what are the numbers?

Let $x$ and $y$ denote the numbers.

From the conditions of the problem,
$$x - y = 2 \quad . \quad . \quad (1)$$
$$x^3y^3 = 42875 \quad . \quad . \quad (2);$$

extracting the cube root of both members of (2),
$$xy = 35; \quad \therefore \quad y = \frac{35}{x};$$

substituting and reducing,
$$x^2 - 2x = 35,$$
$$x = 1 \pm \sqrt{35+1} = 1 \pm 6;$$
$$\therefore \quad x = 7, \text{ and } -5, \quad y = 5, \text{ and } -7.$$

29. A sets out from C towards D, and travels 8 miles each day; after he had gone 27 miles, B sets out from D towards C, and goes each day $\frac{1}{20}$ of the whole distance from D to C; after he had travelled as many days as he goes miles in each day, he met A · what is the distance from D to C?

Let $x$ denote the number of miles from D to C.

Then, $\frac{x}{20}$ will denote the number of miles B travels per day, also the number of days that he travels;

hence, $\frac{x^2}{400}$ denotes the number of miles travelled by B,

$27 + 8x$ " " " " " A.

From the conditions of the problem,

$$\frac{x^2}{400} + 27 + \frac{8x}{20} = x;$$

clearing of fractions and reducing,

$$x^2 - 240x = -10800;$$

$$\therefore x = 120 \pm \sqrt{-10800 + 14400} = 120 \pm 60;$$

whence, $x = 60, \quad x = 180.$

30. There are three numbers; the difference of the differences of the 1st and 2d, and 2d and 3d, is 4; their sum is 40, and their continued product is 1764: what are the numbers?

Let $x$, $y$ and $z$ denote the numbers.

From the conditions of the problem,

$$(x - y) - (y - z) = 4 \quad . \quad . \quad (1)$$
$$x + y + z = 40 \quad . \quad . \quad (2)$$
$$xyz = 1764 \quad . \quad . \quad (3);$$

combining (1) and (2), eliminating $x$ and $z$,

$$3y = 36; \quad \therefore \quad y = 12;$$

substituting in (2) and (3),

$$x + z = 28 \quad . \quad . \quad (4)$$
$$xz = 147 \quad . \quad . \quad (5);$$

combining (4) and (5),

$$x = 7, \text{ or } 21; \quad y = 21, \text{ or } 7.$$

31. There are three numbers in arithmetical progression: the sum of their squares is 93, and if the first be multiplied by 3, the second by 4, and the third by 5, the sum of the products will be 66: what are the numbers?

Let $x$ denote the first number, and $y$ their common difference.

From the conditions of the problem,

$$x^2 + (x + y)^2 + (x + 2y)^2 = 93 \quad . \quad . \quad (1)$$
$$3x + 4(x + y) + 5(x + 2y) = 66 \quad . \quad . \quad (2);$$

performing indicated operations and reducing,

$$3x^2 + 5y^2 + 6xy = 93 \quad . \quad . \quad (3)$$
$$12x + 14y = 66, \text{ or } 6x + 7y = 33 \quad . \quad . \quad (4).$$

From (4), $\quad y = \dfrac{33 - 6x}{7};$

$\therefore \quad y^2 = \dfrac{1089 - 396x + 36x^2}{49},$ and $xy = \dfrac{33x - 6x^2}{7};$

substituting in (3) and reducing,

$$x^2 - \frac{198}{25}x = -\frac{296}{25};$$

whence, $x = \dfrac{99}{25} \pm \sqrt{-\dfrac{296}{25} + \dfrac{9801}{625}} = \dfrac{99 \pm 49}{25}$;

$\therefore \; x = \dfrac{148}{25}, \quad x = 2.$

Taking the second value of $x$, we find $y = 3$, and the numbers are 2, 5 and 8.

The problem supposes the numbers entire, therefore the 1st value of $x$ is not used.

32. There are three numbers in arithmetical progression whose sum is 9, and the sum of their fourth powers is 353; what are the numbers?

Let $x$, $y$ and $z$ denote the numbers.

From the conditions of the problem,

$$2y = x + z \quad . \; . \quad (1)$$
$$x + y + z = 9 \quad . \; . \quad (2)$$
$$x^4 + y^4 + z^4 = 353 \quad . \; . \quad (3).$$

From (1) and (2) we find $y = 3$;

substituting in (2) and (3),

$$x + z = 6 \quad . \; . \quad (4)$$
$$x^4 + z^4 = 272 \quad . \; . \quad (5);$$

raising both members of (4) to the 4th power,

$$x^4 + 4x^3z + 6x^2z^2 + 4xz^3 + z^4 = 1296 \quad . \; . \quad (6);$$

adding equations (5) and (6), member to member, and dividing by 2

$$x^4 + 2x^3z + 3x^2z^2 + 2xz^3 + z^4 = 784 \quad . \; . \quad (7);$$

extracting the square root of both members,

$$x^2 + xz + z^2 = 28 \quad . \; . \quad (8);$$

squaring both members of (4),

$$x^2 + 2xz + z^2 = 36 \quad . \quad . \quad (9);$$

from (8) and (9) we find

$$xz = 8 \quad . \quad . \quad (10);$$

from (4) and (10) we get

$$x = 2, \text{ or } 4; \quad z = 4, \text{ or } 2:$$

hence, the numbers are 2, 3 and 4.

33. How many terms of the arithmetical progression 1, 3, 5, 7, &c., must be added together to produce the 6th power of 12?

The 6th power of 12 is 2985984.

From Art. 175 we have the formula,

$$n = \frac{d - 2a \pm \sqrt{(d - 2a)^2 + 8dS}}{2d}$$

In the present case, $a = 1$, $d = 2$, and $S = 2985984$;

substituting, $n = \dfrac{\sqrt{16 \times 2985984}}{11} = 1728.$

34. The sum of 6 numbers in arithmetical progression is 48; the product of the common difference by the least term is equal to the number of terms: what are the terms of the progression?

Let $x$ denote the 1st term, and $y$ the common difference.

From the conditions of the problem,

$$6x + 15y = 48, \quad xy = 6; \quad . \; . \; y = \frac{6}{x};$$

substituting and reducing,

$$x^2 - 8x = -15;$$

$\therefore \quad x = 4 \pm \sqrt{-15 + 16} = 4 \pm 1$, or $x = 5$, $x = 3$;

whence, $\quad y = \tfrac{6}{5}, \quad y = 2:$

hence, the series is  3 . 5 . 7 . 9 . 11 . 13,

or  5 . $6\frac{1}{5}$ . $7\frac{2}{5}$ . $8\frac{3}{5}$ . $9\frac{4}{5}$ . 11.

35. What is the sum of 10 square numbers whose square roots are in arithmetical progression the least term of which is 3, and the common difference 2?

Let $x$ denote the sum.

The progression of roots is

3 . 5 . 7 . 9 . 11 . 13 . 15 . 17 . 19 . 21,

and the series of squares,

9 . 25 . 49 . 81 . 121 . 169 . 225 . 289 . 361 . 441.

1st order of diffs,   16, 24, 32, 40, &c.,

2d order of diffs,    8, 8, 8, &c.,

3d order of diffs,    0, 0, &c.

From Art. 210, making

$S' = x$, $a = 9$, $n = 10$, $d_1 = 16$, $d_2 = 8$, $d_3 = 0$, &c.

$x = 90 + 45 \times 16 + 120 \times 8 = 1770$.

36. Three numbers are in geometrical progression whose sum is 95, and the sum of their squares is 3325: what are the numbers?

Let $x$, $y$ and $z$ denote the numbers.

From the conditions of the problem,

$$y^2 = xz \quad . \quad . \quad (1)$$
$$x^2 + y^2 + z^2 = 3325 \quad . \quad . \quad (2)$$
$$x + y + z = 95 \quad . \quad . \quad (3);$$

combining (1) and (2),

$$x^2 + 2xz + z^2 = 3325 + xz \quad . \quad . \quad (4);$$

combining (1) and (3),
$$x + \sqrt{xz} + z = 95 \quad . \quad . \quad (5);$$
from (4) and (5),
$$x + z = \sqrt{3325 + xz}$$
$$x + z = 95 - \sqrt{xz};$$
hence, $\quad \sqrt{3325 + xz} = 95 - \sqrt{xz};$

squaring both members,
$$3325 + xz = 9025 - 190\sqrt{xz} + xz;$$
$$\therefore \sqrt{xz} = 30, \text{ or } xz = 900 \quad . \quad . \quad (6);$$
substituting in (5), $\quad x + z = 65 \quad . \quad . \quad (7)$

from (6) and (7), $\quad x = 20$ and 45,
$$y = 45 \text{ and } 20.$$

37. Three numbers are in geometrical progression: the difference of the first and second is 6; that of the second and third is 15: what are the numbers?

Let $x$, $y$ and $z$ denote the numbers.

From the conditions of the problem,
$$y^2 = xz \quad . \quad . \quad (1)$$
$$x - y = -6; \quad \therefore \quad x = y - 6$$
$$y - z = -15; \quad \therefore \quad z = y + 15,$$
and $\quad xz = y^2 + 9y - 90;$

substituting in (1) we find $\quad y = 10;$
$$\therefore \quad x = 4, \text{ and } z = 25.$$

38. There are three numbers in geometrical progression; the sum

of the first and second is 14, and the difference of the second and third is 15 : what are the numbers?

Let $x$, $y$ and $z$ denote the numbers.

From the conditions of the problem,

$$y^2 = xz \quad . \quad . \quad (1)$$
$$x + y = 14; \quad \therefore \ x = 14 - y$$
$$z - y = 15; \quad \therefore \ z = 15 + y,$$

and $\quad xz = 210 - y - y^2;$

substituting in (1), $\quad y^2 + \dfrac{y}{2} = 105;$

$$\therefore \ y = -\dfrac{1}{4} \pm \sqrt{105 + \dfrac{1}{16}} = \dfrac{-1 \pm 41}{4} = 10, \text{ and } -\dfrac{21}{2};$$

taking the 1st value of $y$, we find

$$x = 4, \quad z = 25.$$

39. A, B and C purchase coffee, sugar and tea at the same prices; A pays $11,62½ for 7½ pounds of coffee, 3 pounds of sugar, and 2¼ pounds of tea; B pays $16,25 for 9 pounds of coffee, 7 pounds of sugar, and 3 pounds of tea; C pays $12,25 for 2 pounds of coffee, 5¼ pounds of sugar, and 4 pounds of tea: what is the price of a pound of each?

Let $x$, $y$ and $z$ denote the number of cents that the coffee, sugar and tea cost, respectively.

From the conditions of the problem,

$$7\tfrac{1}{2}x + 3y + 2\tfrac{1}{4}z = 1162\tfrac{1}{2} \ \cdot \ \cdot \ \cdot \ (1)$$
$$9\ x + 7y + 3\ z = 1625 \ \cdot \ \cdot \ (2)$$
$$2\ x + 5\tfrac{1}{2}y + 4\ z = 1225 \ \cdot \ \cdot \ (3);$$

clearing (1) and (3) of fractions,

$$30x + 12y + 9z = 4650 \quad \cdot \quad \cdot \quad (4)$$
$$4x + 11y + 8z = 2450 \quad \cdot \quad \cdot \quad (5).$$

From (2) and (4),

$$3x - 9y = -225, \quad \text{or} \quad x - 3y = -75 \quad \cdot \quad \cdot \quad (6);$$

from (2) and (5),

$$60x + 23y = 5650 \quad \cdot \quad \cdot \quad (7);$$

from (6) and (7), $\quad y = 50$;

by substitution, $\quad x = 75, \quad z = 200.$

40. Divide 100 into 2 such parts that the sum of their square roots shall be 14.

Let $x$ denote the first part.

From the conditions of the problem,

$$\sqrt{x} + \sqrt{100 - x} = 14;$$

squaring both members and reducing,

$$\sqrt{100x - x^2} = 48;$$

squaring both members and reducing,

$$x^2 - 100x = -2304;$$

$$\therefore \quad x = 50 \pm \sqrt{-2304 + 2500} = 50 \pm 14,$$

$$x = 64, \quad \text{and} \quad 36.$$

41. In a certain company there were three times as many gentlemen as ladies; but afterwards 8 gentlemen with their wives went away, and there then remained five times as many gentlemen as ladies: how many gentlemen, and how many ladies were there originally?

Let $3x$ denote the number of gentlemen; then will $x$ denote the number of ladies.

From the conditions of the problem,
$$3x - 8 = 5(x - 8);$$
$$\therefore \quad x = 16, \text{ and } 3x = 48.$$

42. Find two quantities such that their sum, product, and the difference of their squares, shall all be equal to each other.

Let $x$ and $y$ denote the quantities.

From the conditions of the problem,
$$x + y = xy \quad \cdot \quad (1)$$
$$x^2 - y^2 = xy \quad \cdot \quad (2);$$

by division of (2) by (1), we have
$$x - y = 1, \text{ or } x = y + 1;$$

substituting in (1),
$$2y + 1 = y^2 + y, \text{ or } y^2 - y = 1;$$

whence, $\quad y = \dfrac{1}{2} \pm \sqrt{1 + \dfrac{1}{4}}, \text{ or } y = \dfrac{1 \pm \sqrt{5}}{2};$

hence, $\quad x = \dfrac{3 \pm \sqrt{5}}{2}.$

43. A bought 120 pounds of pepper, and as many pounds of ginger, and had one pound of ginger more for a dollar than of pepper; the whole price of the pepper exceeded that of the ginger by 6 dollars: how many pounds of pepper, and how many of ginger had he for a dollar?

Let $x$ denote the number of pounds of pepper for a dollar.

From the conditions of the problem,

$$\frac{120}{x} - \frac{120}{x+1} = 6, \text{ or } x^2 + x = 20;$$

$$\therefore x = -\frac{1}{2} \pm \sqrt{20 + \frac{1}{4}} = \frac{-1 \pm 9}{2}; \text{ hence, } x = 4.$$

The negative value does not conform to the conditions of the special problem.

44. Divide the number 36 into 3 such parts that the second shall exceed the first by 4, and that the sum of their squares shall be equal to 464.

Let $x$, $y$ and $z$ denote the parts.

From the conditions of the problem,

$$x + y + z = 36 \quad \cdot \cdot \quad (1)$$
$$y - x = 4 \quad \cdot \cdot \quad (2)$$
$$x^2 + y^2 + z^2 = 464 \quad \cdot \cdot \quad (3);$$

from (1), $\quad x^2 + 2xy + y^2 = 1296 - 72z + z^2 \quad \cdot \cdot \quad (4);$
from (2), $\quad x^2 - 2xy + y^2 = 16 \quad \cdot \cdot \cdot \cdot \cdot \quad (5);$

adding (4) and (5), member to member,

$$2x^2 + 2y^2 = 1312 - 72z + z^2 \quad \cdot \cdot \quad (6);$$

from (3), $\quad 2x^2 + 2y^2 = 928 - 2z^2 \quad \cdot \cdot \cdot \cdot \quad (7);$

equating the second members and reducing,

$$z^2 - 24z = -128;$$

$$\therefore z = 12 \pm \sqrt{-128 + 144} = 12 \pm 4;$$

hence, $\quad z = 16, \quad z = 8;$

substituting the first value in (1),

$$x + y = 20 \quad \cdot \cdot \quad (8);$$

from (2) and (8),    $y = 12$ and $x = 8$.

**45.** A gentleman divided a sum of money among 4 persons, so that what the first received was $\frac{1}{2}$ that received by the other three, what the second received was $\frac{1}{3}$ that received by the other three; what the third received was $\frac{1}{4}$ that received by the other three, and it was found that the share of the first exceeded that of the last by $14: what did each receive, and what was the whole sum divided?

Let $x$, $y$, $z$ and $w$ denote the number of dollars that each received.

From the conditions of the problem,

$$2x = y + z + w \quad \cdot \cdot \quad (1)$$
$$3y = x + z + w \quad \cdot \cdot \quad (2)$$
$$4z = x + y + w \quad \cdot \cdot \quad (3)$$
$$x - w = 14 \quad \cdot \cdot \quad (4);$$

from (2) and (3),

$$x + w = 3y - z$$
$$x + w = 4z - y; \text{ whence, } 3y - z = 4z - y,$$

or    $4y = 5z$,   $z = \frac{4}{5}y$  $\cdot \cdot$  (5);

from (4),    $w = x - 14$  $\cdot \cdot$  (6);

substituting the values of $w$ and $z$ in (1) and (2),

$$2x = y + \tfrac{4}{5}y + x - 14$$
$$3y = x + \tfrac{4}{5}y + x - 14; \text{ whence, by reduction,}$$
$$5x - 9y = -70$$
$$10x - 11y = 70;$$

$\therefore$   $x = 40$, $y = 30$; and by substitution, $z = 24$, $w = 26$.

**46.** A woman bought a certain number of eggs at 2 for a penny, and as many more at 3 for a penny, but on selling them at the rate of 5 for 2 pence, she lost 4 pence by the bargain; how many did she buy?

Let $x$ denote the number at each price. Then will $\dfrac{x}{2} + \dfrac{x}{3}$ denote the number of pence paid, and $\dfrac{2(x+x)}{5}$ will denote the number of pence received.

From the conditions of the problem,

$$\frac{x}{2} + \frac{x}{3} = \frac{2(x+x)}{5} + 4; \quad \text{reducing,} \quad x = 120.$$

47. Two travellers set out together and travel in the same direction; the first goes 28 miles the first day, 26 the second day, 24 the third day, and so on, travelling 2 miles less each day; the second travels uniformly at the rate of 20 miles a day: in how many days will they be together again?

Let $x$ denote the required number of days. The distance travelled by the first in $x$ days is

[(Art. 176), since $a = 28$, $d = -2$, and $n = x$], denoted by

$$\tfrac{1}{2}x[56 - (x-1)^2], \quad \text{or} \quad 29x - x^2;$$

and the distance travelled by the second is denoted by $20x$: hence, we have

$$29x - x^2 = 20x, \quad \text{or} \quad x = 9.$$

48. A farmer sold to one man 30 bushels of wheat and 40 of barley, for which he received 270 shillings. To a second man he sold 50 bushels of wheat and 30 of barley, at the same prices, and received for them 340 shillings: what was the price of each?

Let $x$ denote the number of shillings for 1 bushel of wheat, and $y$ " " " " " " " barley.

From the conditions of the problem,

$$30x + 40y = 270 \quad \cdot \quad \cdot \quad (1)$$
$$50x + 30y = 340 \quad \cdot \quad \cdot \quad (2);$$

whence, $\quad 110y = 330$, or $y = 3;$ hence, $x = 5$.

49. There are two numbers whose difference is 15, and half their product is equal to the cube of the lesser number: what are the numbers?

Let $x$ and $y$ denote the numbers;

from the conditions of the problem,

$$x - y = 15$$
$$xy = 2y^3 \cdot \qquad \text{or,} \qquad x = 2y^2;$$

substituting and reducing,

$$y^2 - \frac{1}{2}y = \frac{15}{2};$$

$$\therefore \ y = \frac{1}{4} \pm \sqrt{\frac{15}{2} + \frac{1}{16}} = \frac{1 \pm 11}{4};$$

hence, $y = 3$, and $-\dfrac{5}{2}$; also, $x = 18$, and $\dfrac{25}{2}$.

50. A merchant has two barrels and a certain number of gallons of wine in each. In order to have an equal quantity in each, he drew as much out of the first cask into the second as it already contained; then again he drew as much out of the second into the first as it then contained; and lastly, he drew again as much from the first into the second as it then contained, when he found that there was 16 gallons in each cask: how many gallons did each originally contain?

Let $x$ denote the number of gallons in the first cask, and $y$ the number in the second;

$x - y$ will denote the quantity in the first cask after the first drawing, and $2y$ the quantity in the second cask; after the second drawing, $2y - (x - y)$ or $3y - x$ will denote the quantity in the second, and $2x - 2y$ the quantity in the first cask; after the third drawing, $2x - 2y - (3y - x)$ or $3x - 5y$ will denote the quantity in the first cask, and $6y - 2x$ the quantity in the second.

From the conditions of the problem,
$$3x - 5y = 16$$
$$6y - 2x = 16.$$

By combination,
$$x = 22; \qquad y = 10.$$

www.ingramcontent.com/pod-product-compliance
Lightning Source LLC
Chambersburg PA
CBHW020920230426
43666CB00008B/1508